OUR STORY

CELEBRATING THE 20ᵀᴴ ANNIVERSARY OF THE 1998 TITANIC EXPEDITION

OUR STORY

CELEBRATING
THE 20ᵀᴴ ANNIVERSARY
OF THE
1998 TITANIC EXPEDITION

Compiled and Written by Bill Willard

JH
Jomága House

No part of this book may be reproduced or transmitted in any form or by any means, graphic, electronic or mechanical, including photocopying, recording, taping, or by any information storage and retrieval system, without permission in writing from the author except in the case of brief quotations embodied in critical articles and reviews.

Front Cover Photo: The Big Piece before it breaks the surface. Photo courtesy of Bob Sitrick, and RMS Titanic, Inc.

Back Cover Photo: George Tulloch and PH Nargeolet in a press conference at the conclusion of the 1996 Expedition. Courtesy of Günter Bäbler, and RMS Titanic, Inc.

Cover Background: Blue sea surface with waves © leungchopan Design Photos

OUR STORY
   CELEBRATING THE 20TH ANNIVERSARY OF THE 1998 TITANIC EXPEDITION

ISBN: 978-1-60495-041-0

Copyright © 2018 Bill Willard. Published by Jomága House, Broken Arrow, OK. All rights reserved.

CONTENTS

Preface 7
The Leaders 9
The 1996 Expedition — The Adventure Begins 13
Titanic '98 Planning Session – Toulon 23
Preparations for a New Expedition 31
On Our Way 40
Arriving at the Site 56
Recovery of The Big Piece 72
The *Titanic Live* Program 97
Intermission — The First Phase Ends 112
Expedition Photos 119
The Second Phase 155
Expedition Members on the Second Phase 156
The Second Phase Begins 157
Flyover 177
Pieces of the Past 180
Hurricane Bonnie 187
Homeward Bound 202
Reflections... 213
Stories from Expedition Members 221
   Thoughts on the 1998 Expedition - George Tulloch 222
   Memories of *Titanic* – Sara James 225
   From the Old World to the New – John P. Eaton 228
   Dateline's *Raising the Titanic* and *Dateline*/Discovery
     Special *Titanic Live* – Bob McKeown 237
   Memories – Bob Reid 247
   Titanic Memories – Angus Best 251
   Illuminating History – Christian Petron 256
Honored Friends 261
Author's Notes 272

Preface

The 1998 Titanic Expedition brought the best of many worlds together for a brief time — a compressed time — to attempt a broad scope of objectives on the bottom of the ocean, on ships and in studios throughout the world. Led by visionary George Tulloch, diverse components worked together to complete an immense project. It is this project and these people we wish to salute as we celebrate the 20th anniversary of the adventure when we raised The Big Piece and produced a documentary aired live, which would be the second-most-watched live event ever.

Many expedition members have shared their memories to produce this commemoration of extraordinary efforts, all designed to add more to the story of *Titanic*, to a knowledge of the ship, of her passengers and crew, and to the events of those fateful hours so long ago. It is not only for them, but also for generations to come that we accomplished the many objectives set before us.

This is our story.

The Leaders

RMS Titanic Inc.

    George Tulloch — president, expedition leader — "George"

    Paul-Henri ("PH") Nargeolet — co-expedition leader, former member of IFREMER, Lt. Commander retired, French Navy

The Discovery Channel, Inc.

    Maureen Lemire — executive producer

    Bob Sitrick — live production specialist

NBC

    Meade Jorgensen — executive producer

    Jack Bennett — technical manager

    Bob McKeown — anchor reporter from *20/20*

    Sara James — anchor reporter from *20/20*

IFREMER — L'Institut Français de Recherche pour l'Exploitation de la Mer – (The French Institute for Research and Exploitation of the Sea)

    Pierre Valdy — project lead

    Jean-Michel Nivaggioli — chief of manned submersibles department

Stardust Visuals — team that produced the video programming for the Discovery Channel

    Greg Andorfer — executive producer

    Gary Hines — producer

    David Elisco — producer

    Charlene Haislip — coordinating producer

    Saul Rouda — sound technician

Our Story

    Mark Knobil — cameraman

    Brandon Plonka — production assistant

Historians

    Günter Bäbler

    Steven Biel

    John P. Eaton

    Charles A. Haas

    Claes-Göran Wetterholm

Scientists

    Angus Best

    Dr. Roy Cullimore

    William "Bill" Garzke

    Lori Johnston

    Paul Matthias — Polaris Imaging

    Dick Silloway

    Tim Weihs

    David Wood

Oceaneering, Inc.

    George Brotchi

    Troy Launay, pilot

    Ron Schmidt, pilot

Nauticos

    Tom Dettweiller

Others

    Susan Wels — author who journaled the daily happenings and her personal experience

David Livingstone — chief architect, Harland and Wolff

Christian Petron — underwater filming and diving expert

THE CREWS OF:

Ocean Voyager — expedition ship, center of all NBC operations, video capture and production

Nadir — IFREMER vessel, home of *Nautile*, conservation base

Abeille Supporter — vessel recovers Big Piece, home for *Hysub* ROV

Petrel V — taxi ship, used as extra sleeping quarters while on site

Nautile — yellow, manned submersible of IFREMER, carrying two passengers and a pilot

Robin – small remotely operated vehicle (ROV) capable of penetration into the interiors of *Titanic*

Remora — Phoenix International ROV

T-Rex — the mini-ROV designed and operated by Bill Willard and Susan Willard

RMS Titanic, Inc., led by expedition leaders George Tulloch and PH Nargeolet, coordinated efforts and objectives of the undersea components of the expedition and operated from the research vessel *Ocean Voyager* (known as *OV* to everyone), and *Nadir*. RMSTI maintains the legal rights to salvage artifacts from the *Titanic* site in compliance with the Admiralty Court (Norfolk, Virginia).

The French team from IFREMER, with Pierre Valdy as a project leader, brought the submersible *Nautile* onboard the research vessel *Nadir*. The French team members were highly respected veterans of oceanographic exploration.

The Discovery Channel co-sponsored the expedition with Executive Producer Maureen Lemire at the helm. Also co-sponsoring was NBC, with Executive Producer Meade Jorgensen coordinating the NBC programming and Jack Bennett managing the technical operations. Stardust Visuals produced the documentary shows aired after the expedition. Bob Sitrick coordinated the *Titanic Live* program aired on six different continents.

Two additional vessels participated: the *Abeille Supporter*, which recovered and transported The Big Piece, and the *Petrel V* — as one expedition member said, "at least it floated."

Also on the expedition team were historians, authors, scientists, forensic analysts, camera and sound technicians, and ROV teams for Oceaneering's *Magellan* ROV, the Phoenix *Remora* ROV, and the *T-Rex* ROV. Specialists in many areas were included.

There were so many brilliant and inspiring people associated with this expedition that it is difficult and perhaps unfair to single out a few.

Bob Sitrick

~ 1 ~

The 1996 Expedition: The Adventure Begins

The 1998 expedition developed from unfulfilled goals during the 1996 expedition. George Tulloch had planned for RMSTI to conduct expeditions at the wreck site every two to three years. His goals were to perform site surveys, to explore the wreck historically and scientifically, and to recover selected artifacts for conserving and displaying in a series of traveling themed exhibitions.

The crew of the '96 expedition was an experienced group. IFREMER would assist in the recovery of a large, detached section of the hull, estimated to weigh 20 tons and affectionately named "The Big Piece."

> Every time I look at it, it seems bigger . . . . Very big. I don't know about this.
>
> PH Nargeolet

The team also included noted historians and technical analysts, namely David Livingstone, chief architect at the Harland and Wolff shipyard in Belfast, and Bill Garzke, Naval forensics expert.

> We had been part of the 1996 Titanic Expedition to locate and retrieve a large structural section of *Titanic*. One of our tasks on that voyage was to identify where

this piece came from on the ship, how much it weighed, and find its location on the seabed. One of the principle tasks of this expedition was to bring this structural section of *Titanic* to the surface so that it could be put on display as part of the RMS Titanic Exhibit.

Bill Garzke

Research of the rivet pattern on the hull plating estimated the piece to be approximately 27 feet long. Historians studied the porthole pattern and gave a preliminary location of where this section had been part of the ship and identified staterooms on the interior side of the section.

Pierre Valdy of IFREMER, partners with RMSTI in '96, was project leader. Valdy's plan called for the piece to be secured to seven lift bags, and, at the appropriate time, brought to the ocean's surface. It was a well-thought-out plan. A lift bag was a large Kevlar bag filled with 5,000 gallons of diesel fuel. Diesel fuel is non-compressible at *Titanic*'s depth, ultimately giving one bag a lift capacity of 7,000 pounds — three-and-a-half tons. When recovered at the surface, the fuel would be returned to the ship's fuel tank. The procedure was simple: filled lift bags would be attached to several tons of scrap iron on the ship's deck. When the scrap iron was displaced overboard, the iron weight, or ballast, would outweigh the buoyant force, causing the lift bags to be pulled to the ocean floor. There, *Nautile*'s manipulator arms would attach the lift bags to designated locations on The Big Piece. A signal from the surface when all bags were in place would sever the lines to the ballast iron, and the bags, now attached only to The Big Piece, would combine for more than 24 tons of lift. The depths involved, plus the large volume of diesel fuel to transfer meant that it was a slow preparation process over several days.

There is no greater legend or mystery of the sea than *Titanic*. This mission is primary in our lives. So you can imagine the joy that will be reflected in our faces when we see the ship's hull re-emerge from the ocean that claimed it 84 years ago.

George Tulloch

The '96 expedition had a second objective. RMSTI partnered with the Suarez Corporation for a cruise ship to bring paying passengers to witness the recovery. Interest was significant. The first ship began to sell cabins quickly so a second ship was chartered, doubling the number of enthusiasts who wanted to witness the recovery of The Big Piece.

The addition of the cruise ships originally was a flash of brilliance for public relations, but the situation quickly evolved into a distraction of major proportions. The two cruise ships on site, the *Royal Majesty* and the *Island Breeze*, held 1,700 paying passengers, historians, and celebrities, including a few *Titanic* survivors. A large number of passengers on both ships alleged that in the promotional materials they received while booking their cabins, they were promised Zodiac rides and tours of the expedition ships.

George Tulloch was pulled in two directions, wanting to move the expedition on schedule but also trying to satisfy the passengers. Zodiac rides were out of the question, per the captains of both cruise vessels, due to the extreme risk to passengers under their charge. The passengers were upset and would not be appeased even though experts and celebrities moved from ship to ship for personal appearances and lectures. Book signings, talks, and socials happened daily. *Titanic* survivors Eleanor Johnson Shuman, Edith Brown Haisman, and Michel Navratil signed autographs and posed for photos between talks. William McQuitty, producer of the film *A Night to Remember*, astronaut Buzz Aldrin and actress Loni Anderson also entertained passengers. Ann Lightoller, a great-granddaughter of *Titanic*'s Second

Our Story

Officer C.H. Lightoller shared family stories.

As advertised, the expedition highlights were broadcast from Ocean *Voyager* (*OV*) to both cruise ships. Passengers could watch segments prepared just for them on all ship televisions, even in their rooms, if desired. Host Bob Anderson was tasked with narrating the shows.

> I provided daily live broadcasts from the 1996 Titanic Expedition research vessel *Nadir* plus daily discoveries, historical and scientific reviews and personal stories that connected the past to the present. We were all preparing with the understanding that a special celebrity host, Alan Alda, would be arriving on the first cruise ship to fulfill the role as the on-camera host/presenter talent. However, as that first ship approached (contact was established at about 28 miles out) and the planned technical check was underway, one of the event cruise directors requested a word with the expedition leader, George Tulloch, to inform him that "Host" Alan Alda did not make the trip!!! The brief and rather abrupt discussion between George and me determined that due to his duties for personal appearances on multiple vessels, he could not host the broadcast. As he leaped into the waiting Zodiac (our inflatable sea taxi) he simply said, "Take care of it Bobby." And off they sped toward the horizon and beyond.
>
> Bob Anderson

Bob had a short time to pull everything together and step into his new role. Meanwhile, lift bags were deployed to the ocean floor, and preparations continued for the recovery. Over the next dives, the *Nautile* team secured the lift bags to The Big Piece. Arriving on time at the site was the vessel *Jim Kilabuk*, tasked with lifting The Big Piece out of the water.

Thursday, August 29, 1996

On the two cruise ships, expectant passengers lined the rails as the ships circled the expedition vessels. It was a hot day with blue skies. Bob Anderson's broadcast informed the observers of the mission's progress.

> Only a few onboard the *Royal Majesty* knew the story of *Titanic* well enough and knew about the nature of an expedition (at least in theory), so I was busy explaining what was going on, etc. People thought The Big Piece might suddenly float on the surface, so the very basics, like "no lift bags – no Big Piece" were shared countless times.
>
> Günter Bäbler

The excitement of the passengers caused confusion among the cruise employees. The casino croupiers were left standing at their tables with only a handful of players. The dining hall attendants had no guests to serve. The lounges had very few clients. Creative thinking gave lounge managers an idea, and they sent their servers out to the people on the deck, taking orders then serving beverages to a thirsty crowd. Though being told from the Anderson broadcast that the lift bags had a travel time of at least two-and-a-half hours, some passengers were expecting a much faster arrival.

Anderson finally made the announcement — The Big Piece was in motion and on its way upward. Two hours, however, seemed an eternity for impatient enthusiasts wanting to see a doubly-historical event. Not only would they see large part of a sunken wreck being recovered by an expedition team, but for the first time in 84 years, they would view a portion of the hull of the greatest ship ever built — *Titanic*.

Our Story

Thursday, August 29, 3:15 p.m.

The two cruise ships continued to circle the expedition vessels. Suddenly someone pointed, another person shouted, and cameras began to photograph the first lift bag bobbing on the surface. The anticipation intensified, and the atmosphere became electric. Two more lift bags appeared, and The Big Piece arrived to cheers, shouts of excitement and a visibly joyous crowd of nearly 1,700 observers.

The expedition team began Phase Two — getting The Big Piece from the water onto the deck of the recovery ship, *Jim Kilabuk*. Everyone watched with eagerness as the *Kilabuk* moved toward the lift bags, positioning for the recovery.

> I looked at the *Jim Kilabuk*, and even today I am convinced that The Big Piece would have never fit under the A-Frame – no way. The vessel was not suitable for the mission.
>
> Günter Bäbler

The winch aboard the *Kilabuk* was inadequate; it could not lift The Big Piece from the water. A year's worth of plans lay beneath several lift bags, two and one-half miles above the ocean floor, with no way to pull it the final 200 feet onto the deck of the recovery ship.

The lengthy delays confused many of those watching. The expedition team could not communicate the complex problem right away. Slowly, passengers began to disperse; many groups went to dinner, many decided to relax and cool down after being in the sun all afternoon. Many just sat, hoping for the expedition team to engineer a way to get The Big Piece out of the water. A small group began to complain.

> The emotions of the passengers were constantly changing from emotionally up to emotionally down while at the wreck site, since the news on what was

happening spread slowly. Some passengers were even offered a chance to sign up for a Zodiac ride to the *Nadir*. Several were so upset, they were seeking fellow passengers to sue RMS Titanic Inc. for not keeping promises, but clearly these people came on board with wrong and unrealistic expectations.

Günter Bäbler

In the midst of this dilemma, a new, unexpected piece of the puzzle entered the picture. Hurricane Edouard had departed the eastern seaboard and was headed toward the expedition site. Now, the team faced two challenges. The first was recovering the 20-ton piece of *Titanic*'s hull now hanging beneath several lift bags within a few hundred yards from the surface. The second was managing the recovery with the impending arrival of a storm with significant wave swells.

The expedition team ended up with two possible solutions. One option, submitted by a passenger with a physics background, called for The Big Piece to stay attached to the lift bags with transponders, and to allow the Piece to freely ride the swells of the hurricane. The transponders would keep track of the Piece's location at all times. When the hurricane's effects passed, the *Kilabuk* would secure the Piece then tow it toward Halifax. At that time, a ship with an appropriate lifting winch could be dispatched. The Big Piece would be recovered where the ocean depth would be 300 feet.

The second option, proposed by the captain of the *Kilabuk*, would be to detach the lift bags after attaching the Piece to the *Kilabuk*, and drag it toward Halifax. The expedition leaders chose the second option. The Big Piece was secured to the *Kilabuk*, the lift bags were detached, and slowly the *Kilabuk* sailed toward Halifax. The cruise ships turned toward their home ports, New York and Boston.

Our Story

Friday, August 30, 1996, 2:30 a.m.

During the evening, the intensity of the waves increased, as did their heights. The Big Piece dangled, in essence, a heavy iron weight beneath the *Kilabuk's* stern, causing distress on the ship. Twenty tons bobbing with the waves could easily capsize the ship or drag it down beneath the surface. The expedition documentary stated several supporting cables attaching The Big Piece to the lift bags broke, and The Big Piece fell to the ocean floor.

Later in the day, when calmer seas returned, *Nadir* launched *Nautile*, and The Big Piece was located, sitting upright, wedged into the sediment. The submersible crew spent several hours ascertaining its condition. A photograph soon appeared with a message, handwritten on a metal tag near the porthole along with a locator transponder: "I will come back – George Tulloch." It was a wonderful quote, expressing hope that RMSTI would be back for The Big Piece, but the caption was not written by George Tulloch. This incident has become one of many legends in company lore.

The '96 expedition concluded, accomplishing many objectives, yet with a reminder that nature does not always cooperate, and that even the best made plans sometimes have glitches. As the expedition team headed home, they did so with an unfulfilled dream, and planning would begin right away on a second attempt to raise The Big Piece.

What happened to The Big Piece during the hurricane has always been a point that RMSTI did not openly discuss. It is time for the complete story.

Prior to the '96 expedition, an agreement was reached between the owner of the *Jim Kilabuk* and RMSTI that the ship's A-frame was not adequate for the recovery, and that a crane would be needed. PH Nargeolet himself inspected a crane belonging to a third party. He sought and received approval from the Canadian Coast

The 1996 Expedition: The Adventure Begins

Guard for the crane to be used in the recovery. Approval by the Coast Guard meant the crane was in compliance with regulations and satisfied insurance requirements. However, when the *Kilabuk* arrived on site for the recovery, the A-frame was prominent, but there was no crane. The person on the *Kilabuk* in charge of the recovery had made several critical errors which doomed the attempt.

> The guy in charge of loading the piece was incompetent. The back roll was locked, and it was impossible to unlock it. That is why some of the lines broke.
>
> PH Nargeolet

It was not known why the *Kilabuk's* officer did not bring the crane as agreed. The back roll, sometimes called a stern roller, is designed to act like a pulley, allowing cables over it to slide freely as the tide rises and lowers. By locking the roller, the ship's movement increased the tension in the cables.

> He pulled on the main cables until they broke.
>
> PH Nargeolet

When the choice was made to tow The Big Piece, it was discovered that the current was moving in the opposite direction from what the recovery officer had told them. This current would make the towing process almost impossible. At one time, the *Kilabuk* went into reverse — creating a high risk of severing the tow ropes.

> The *Kilabuk* was doing things so stupid that we asked the ship to release The Big Piece.
>
> PH Nargeolet

A frustrated RMST leadership made the decision. The towing lines of The Big Piece were transferred to *Nadir*, and as the storm intensified, one lift bag, and soon a second lift bag separated due to the stress. George and PH asked *Nadir's* captain about options. It was agreed that it would be impossible to lift The Big Piece out of the water and secure it on the fantail during the storm. It was unsafe to drag The Big Piece; allowing it to float freely wouldn't work since the lift bags had detached. As heart wrenching as it was, George and PH knew the only practical option was to release it, locate it on the ocean floor, and return to complete the mission at a later date.

> This is the true story about The Big Piece and the *Kilabuk*. For years George and I didn't want [to talk about] the story. Now, it is time to tell it.
>
> PH Nargeolet

As an epilogue to the *Kilabuk* story, RMST cited breach of the agreement and did not pay the company that owned the *Kilbuk*. The owners, on hearing the issues, immediately terminated the officer in charge of the recovery.

Charles Haas, historian and author who was also a member of this expedition, summed up the '96 attempt perfectly:

> In retrospect, it would have been so much easier for George and PH to have dispensed with public participation at the site [during the recovery attempt]. The Big Piece's loss could have been kept a secret. But these two men and their dedicated colleagues believed in letting all the world see their work. In doing so, they gave everyone aboard all six vessels unforgettable lessons in persistence and courage that armchair critics and myopic journalists couldn't even conceive.... The Big Piece may not have surfaced during the Titanic expedition of 1996, but so much more did.
>
> Charles A. Haas

~ 2 ~

Titanic '98 Planning Session Toulon

The magnitude of the next expedition increased during the year following the '96 expedition. Ratings from the Discovery shows from '96 illustrated that there was an audience thirsting for anything "*Titanic*." Exhibitions at Nauticus in Norfolk, Virginia and The Pyramid in Memphis, Tennessee drew large crowds. The exhibit at the Florida International Museum in St. Petersburg, Florida garnered more than 51,800 advance reservations from an eager public in October '97. James Cameron's film *Titanic* was released in December; both young and old were introduced to a story that for many years seemed to be only a legend.

To say that George Tulloch was a dreamer would be an understatement. His vision, coupled with his unending energy and enthusiasm, motivated him. He reached out to new entities to accomplish his goal. New partners in the project, to accompany Discovery, Stardust Visuals, and IFREMER, were NBC and numerous historians and scientists.

Our Story

May 1998, Toulon, France

Shortly after the 1996 Expedition, at the Nauticus Maritime Science Center in Norfolk, I attended the second day of the artifact exhibit there. PH and George came into the exhibit, and PH recognized me from months earlier. We began to talk near the gift shop. I inquired about the depths to which *Robin*, the IFREMER ROV had explored. PH described some of the complications within the bow of *Titanic* – wiring, blocked passages, navigational difficulties and more. I asked him simply, "If I design a small expendable ROV, would it be helpful?" As we talked, a small crowd began to gather around us, listening. We exchanged business cards after realizing our conversation had lasted the better part of a half hour. Over the course of the next months we continued to correspond and discuss the possibilities. George invited our family to be a part of the 1998 Expedition, with our family-designed and built mini-ROV, *T-Rex*, designed small and light for entry into the most limited areas of the ship.

In April, I received a call from one of the Discovery Channel executives, informing me of the planning session in Toulon, France in early May and requesting my attendance at that planning session. I had three weeks to get a passport and a plane ticket. When I landed at the Nice airport, I was scheduled to meet with three others, and we would take one car to Toulon. Gathering my luggage, I turned toward the exit.

Standing off to the side was a plainly dressed gentleman, wearing glasses and holding a sign that said "Bill Willard." I approached with a smile on my face. "Bon jour! That is me!" I said, pointing to the sign, then to myself. He responded with mumbles, shrugging his shoulders as if he did not understand. Remembering my first semester French, I countered, "Je m'appelle

Bill Willard." I pointed again to the sign, then to myself. Again, the man shrugged his shoulders shaking his head, not understanding. My attempts continued, trying to be more and more basic, each time without success. The mumbled phrases became longer and more incoherent, the puzzled look became more and more pronounced, and I reached the end of the proverbial rope. "Well friend, I have no other ways to tell you that I am Bill Willard. I can't understand a bit of French."

"Neither can I," the man responded in perfect English. "Let's go to Toulon." He grinned a wide grin, and pointed to a bench where Gary Hines and Charlene Haislip, the other members of our party, were laughing out loud! "I'm Bob Anderson! Glad to meet you!"

Bill Willard

One of the objectives for '98 was to produce a first-time-ever live show from the wreck site, airing real-time images from *Titanic* around the world.

A new unmanned robotic system was recruited; Travocean's *Hysub* would be aboard the support vessel the *Abeille Supporter*.

The goal was to broadcast live not only from the middle of the ocean, and not only from an unmanned submersible at the bottom of the sea, but [also] from a manned submersible that could not (for safety reasons) be within two-and-a-half miles of the unmanned sub and its tether. The fear was that it if the cable snapped at the surface it could trap the manned sub at the bottom. The brilliant scientists in France came up with a fiber optic cable that could be attached to the unmanned sub and left at the bottom so that the manned sub could swim up, connect itself with the robotic arm, and broadcast live through the unmanned sub. There were only two of these connectors – one for test day

(day before live show) and one for live day. The result was the greatest goose-bump moment of my life – going on the air and having our announcer say, "You are looking at something no-one has ever seen before – the first live images from a manned submersible at the bottom of the ocean and site of the *Titanic*."

Bob Sitrick

At the meeting tables in Toulon sat a prestigious group of dreamers and planners. There was Jean-Louis Michel, who was involved in finding *Titanic* along with Robert Ballard. Beside Michel sat Tom Dettweiler and Bruce Brown from Nauticos, enlisted by Discovery to oversee operations. Tom Dettweiler was at the monitor with Michel and Ballard the night *Titanic* was found. Meade Jorgensen and Jack Bennett were there from NBC, and across from them were the Discovery representatives led by Maureen Lemire. Continuing around the table, there sat Bob Sitrick, a specialist in live productions, followed by George Tulloch, who was continually up and about, moving from group to group. PH Nargeolet sat next to George, and the two huddled often, with George's animated hands showing his excitement. The IFREMER team sat next to PH, and often, PH and Pierre Valdy would have private discussions. The Stardust representatives were next to the IFREMER team. Bob Anderson and Bill Willard completed the group.

Topics were discussed based on an agenda, and the two sponsoring groups – Discovery and NBC – would talk, then break off to discuss among themselves, make calls to the corporate offices, then return to the table. This procedure happened on numerous points. Each new objective point was discussed for safety (IFREMER was uncompromising on this issue), practicality, and cost. Each group wanted to negotiate who would cover which costs. By the middle of the afternoon, when one discussion point became a bit tenuous, George interrupted and called for a half-hour recess.

Pierre Valdy looked over at me, and with his contagious smile, motioned for me and two others to follow him. We walked out the back of the building, and docked there was the *Nadir*, with *Nautile* on deck. "Want to go inside *Nautile*?" he asked. He did not have to ask twice. One at a time, the three of us climbed down into the submersible. I was surprised at the limited space inside. I stood over six feet tall, so I had to bend over significantly while moving around inside. I laid on an observation pad, looking out the window as one would while viewing *Titanic*. The controls were simple, and translating the labels was an easy task. Pierre very graciously answered our questions.

Bill Willard

At the end of the break, everyone returned to the meeting room to continue the planning session.

If goal #1 wasn't ambitious enough, goal #2 was to raise a 20-ton piece of the *Titanic* hull to the surface – what is known as The Big Piece. The ingenious strategy that the French devised for this challenge was to utilize lift bags. The bags were dropped precisely to the sea floor using weights – then affixed to The Big Piece at the bottom using the unmanned submersibles' robotic arms, which they [used to] cut the weights loose and send the lift bags and cargo to the surface.

Bob Sitrick

Pierre Valdy explained succinctly and thoroughly his lift bag procedures, and proposed a timeline for lifting The Big Piece. A comment from the table, "It's coming up this time!" was followed by a mix of cheers and clapping.

Our Story

The excitement began to outweigh the stress of planning.

Having designed and constructed the ROV *T-Rex*, Bill Willard explained its purpose and answered questions. As the last question was answered, George jumped up, quickly came to the front of the room, and praised everyone's efforts. The team excitedly congratulated one another with energetic handshakes and friendly hugs.

That first evening, the team dined at a restaurant on the coast, overlooking the bay. Christian Petron served as host, telling of several excellent diving targets in the bay, including a German Messerschmitt airplane shot down during World War II. Christian interpreted the menu for those with little or no French. The meal was delicious, and the company and conversation were magnificent.

Dinner was planned for another wonderful restaurant the second night. A large room was reserved for the party, and conversations continued as everyone gathered. One of the situations discussed at this dinner was how to integrate the *T-Rex* system with the mother ROV, the *Magellan*. It was felt that Nauticos would be the quickest route to interface the *T-Rex* controls through the auxiliary controls of *Magellan*, basically, to piggyback the *T-Rex* signals through the *Magellan* system. Plans would be finalized in the next few months.

Following the meal, one of the team members experienced a series of extraordinary events.

> I was approached by someone handling the session's logistics. Since my plane was the first to depart Nice Airport – an hour away – I was asked if I would be willing to take the train into Nice that night to spend the night across from the airport. I agreed, and we began the process.
>
> I boarded a train in Toulon with Florence Nargeolet (PH's wife) explaining to the others in the compartment that I spoke little French, and for them to make sure I

made it safely to Nice. It was nice, in basic English and French, sharing the ride with new friends. "Star Wars!" said one. "I like baseball and the New York Yankees!" said another, repeating the English phrases they had learned. I smiled, and responded each time, knowing they had no idea what I was saying.

A problem developed on the train ride. Several – no, more than several – cows had wandered onto the tracks. The train had to stop while people tried to herd the cattle away. The delay was approximately an hour. Our final arrival time in Nice was after midnight, well after the 11:00 closing time of the station. As we walked out, gates were closing behind us. Exiting the station, everyone was picked up but me, and a taxi sat at the curb. Traveler's checks had been advised for this trip, and I had a quantity with me. However, taxis in Nice, France do not accept traveler's checks after banking hours for fear of fraud. The taxi pulled off, and I stood alone on the street.

A bus map was posted across the street. I had two sports bags for luggage, so I put them over my shoulders and walked to the map. Pulling out paper and a pen, I sketched what I thought would be the best and most direct route to the Nice airport. I began my journey on foot, knowing my destination was approximately five miles away. Along the way, something totally unexpected happened. My "direct route" ended up taking me through the red-light district. My conservative upbringing was NOT ready for all that I saw! Fortunately, I did know "Non, merci . . . au revoir" and used it many times!

I finally arrived at the Nice Airport slightly before 4 A.M. and camped beside the door. The airport opened at 5 A.M., and with my flight departing at 7:45, I saw no point in checking into a hotel for just a few hours. This decision caused yet another problem. The Discovery people called the hotel to learn I had never checked in. They did not know of my whereabouts, and the airline would not confirm I had made it to the flight. After two flights and a short drive, I returned home. Shortly after my arrival, the phone rang with a call from France. They were relieved I had made it home safely, and listened intently to my very interesting travel tale.

Bill Willard

The dynamics of the planning meeting had been incredible. Present were two major sponsors, each with a unique piece of the puzzle, each offering something to the project no one else could supply. Each wanted components for their respective shows, and clearly, some components would have significantly more appeal than others. It became a true negotiation. The "who pays for what" discussion was important, and it would become critical later as Hurricane Bonnie approached.

Discussions concerning the *T-Rex* ROV interfacing with *Magellan* continued after Toulon, and a date was scheduled for the Willards to deliver their ROV to the Nauticos offices in Maryland. Upon arrival, Tom Dettweiler and Maureen Lemire greeted the Willards. In addition to seeing the ROV and reviewing control documentation, the group viewed a video of *T-Rex* exploring a sunken vessel. The film had two parts: footage of the vessel from the ROV's camera, and shots of *T-Rex*'s motions captured by a diver, Captain William Routh, who had proposed the dive. *T-Rex* navigated through small openings and around obstacles just as it might encounter at *Titanic*. The full zoom capabilities of the camera system were shown. When the video ended, there was silence in the boardroom. Tom Dettweiler said, "This is impressive. We are going to make this work!"

For many, the serious work was just beginning.

~ 3 ~

Preparations for a New Expedition

After Toulon, the leaders of the various teams involved coordinated an overall plan. Charlene Haislip acted as central communications, gathering the plans and disseminating details to everyone. Every few nights, fax machines would come alive with several pages of planning, "heads and beds" and needed information. As the sailing date drew near, faxes were sent more often. These communications were necessary and appreciated, and many questions were answered in those memoranda. Everyone learned there were four ships departing on four different dates from four different locations. More importantly, each person knew their assignment.

July 26, 1998

In Boston Harbor, *Ocean Voyager* (*OV*) took on two cargo trailers of NBC equipment. A large crane lifted the trailers up and over to the deck, and welders attached them to the upper deck. Several of the workers complained in a perfect "Bahstan" accent. The trailers had been welded in the wrong location, so the team had to release the welds, re-position the trailers, and make a second attempt to secure the stations. Their work was behind schedule by more than a day, and they were not happy.

> As I watched the welders, torches going full speed, I listened to the others in the crew, grinning as they shared their colorful metaphors. I learned we were at the "shipyahd, noth of the pahkway."
>
> Bill Willard

A breezeway separated the forward and aft sections of the ship. The higher decks contained passenger cabins furnished with bunk beds and two small bureaus for clothes and a small desk between them.

The middle decks contained work areas and the lower decks were crew areas. Aft was a fantail, a flat deck close to the ocean that would serve as the launch and recovery area of the ROVs, with a large mechanical crane attached at center aft. A large cabling deployment system was housed off the fantail, with an umbilical that would enable the large Oceaneering ROV *Magellan* to operate at a depth of 12,500 feet. A work area to the port side was the *T-Rex* lab, and farther forward was a large work area for video monitors and recording stations, and a room to be used for editing.

> Pierre, one of the stewards, gave me a quick tour of the dining area and showed the quickest way to the bridge. His heavy French accent was overshadowed by his friendliness and his wide smile as he told me, "How you say . . . make yourself home!"
>
> Bill Willard

The dining area, or galley, was a rectangular room with booths and tables, and had the appearance of an old diner. Upon entering, everyone always reviewed the white board on the wall listing the day's meal options. Lunch and dinner every day offered a soup, an appetizer, entrée and dessert. The meals were prepared in the kitchens below, and sent up to the galley in a dumbwaiter. The two galley stewards served the tables. There were staples and condiments in shelving units, and along the port wall was a refrigerator with additional items. Each entrée included rice, twice every day, served as a mound on the plate, as if formed in a ramekin mold.

The bridge held the main meeting area where all the expedition leaders discussed plans for the day, plus any new topics that needed their attention. This area became the brain center of *Ocean Voyager*.

July 27, 1998

Early in the morning, the NBC team arrived, and the dock crane began to transfer several pallets of secured equipment and videotapes to the upper deck. Several team members joined in NBC's bucket brigade to take all of the equipment and tapes to be stored in the editing room. Later, pallets containing the monitors and electronics were transferred to the ship. By late afternoon, the NBC team had the equipment in place, ready for operations.

July 28, 1998

Other people began to arrive over the next 24 hours. The work on *OV* was completed, and the ship was moved to the World Trade Center area across Boston Harbor. The ship was docked across from the building that housed the Titanic Artifact Exhibition.

> I took an airplane flight to Boston on Tuesday, July 28 to board *Ocean Voyager*. When I reached the ship, I met Dave Livingstone. Greg Andorfer arrived and immediately greeted us and then discussed our duties during the expedition.
>
> Bill Garzke

Our Story

Like Thomas Andrews during the construction and launching of *Titanic*, David Livingstone held the position of chief architect at Harland and Wolff.

> On Tuesday, July 28, David Livingstone had problems in his scheduled flight from Belfast, Northern Ireland. A fire in one of the plane's engines forced him to take another flight, but in doing so, his luggage had to be put on another flight to Boston. After a long day, a very anxious David finally received his luggage. He was finally relieved when the luggage was delivered to *Ocean Voyager* that evening.
>
> Bill Garzke

Being docked across from the exhibit, *OV* was an "extra" for visitors viewing the artifacts. They were told about the expedition, and many walked 100 yards onto the dock to view the ship.

> I had no responsibilities for the day, and found my way to the exhibit. I was given a VIP badge, and introduced as an expedition member. The entire morning and early afternoon was spent answering questions, sharing mission objectives, and praising the collaborative effort of NBC and the Discovery Channel. It was there I met Susan Wels, author of *Titanic: Legacy of the World's Greatest Ocean Liner*. Her book was in the gift shop, so I pointed her out to the people standing nearby: "That lady over there is Susan Wels, the author of this wonderful book. If you decide to purchase this, be sure to have her autograph it for you!"
>
> Bill Willard

On Thursday, July 30, Dave Livingstone and I took a walk into downtown Boston to locate a store that sold beer. Dave wanted beer, but in Massachusetts, alcoholic beverages are sold in package stores. We finally found a store, and while purchasing our meager supplies, we met Maureen Lemire, another expedition member. She and two of her film crew were purchasing wine and other liquors for the voyage. They would rent a Lincoln town car and bring the party of revelries back to the ship with their liquor cargo. Upon arrival at the pier, we learned that this liquor cargo could not be brought aboard ship. *Ocean Voyager,* by agreement with the customs authorities, was to be a dry ship – unknown to us. Nevertheless, we all did manage to find a way to smuggle our purchases aboard. Dave and I were told that they needed to hide one of the wine cases in our stateroom. That allotment of wine would serve a useful purpose later in the voyage.

Bill Garzke

Before we left Boston, I sat in the office area and Tom Dettweiler walked in. Very soon afterwards, we were joined by Maureen Lemire, Charlene Haislip and several others. Tom had a letter in his hands. The expedition team somehow had been contacted by a group similar to the Make-A-Wish Foundation. A young boy was suffering from a terminal illness, and his request was to be a part of our trip. Hearing the request was gut wrenching for most of us. After Tom finished telling us the request, the group sat in complete silence. After a short while, Tom spoke softly, "I don't see how we can honor this." His main concern was that the ship would be at the wreck site for a planned first leg of 18 days. "What happens if this young man needs medical attention? We don't have anything above a first aid kit on board." There were other valid concerns presented. Someone said, "There are alternatives." "Such as?" Tom asked. Someone took suggestions down on paper: Bring the young boy to Boston for a special, private visit with The Big Piece on its return, with a private tour of the exhibit; ask several of the important people on the expedition (George, PH, the television anchors) to attend, and spend a while with the young boy, send him the VHS series of all the programs we were going to produce. Everyone liked those ideas, and more. He would get a package of Discovery items – an expedition hat and shirt, and more. Tom liked this, as did the group; we saw it as an alternative to an expedition on an unpredictable ocean. Tom left us to go contact the group with the alternative offer. I never heard whether the young boy accepted the offer, or made another wish.

Bill Willard

On the afternoon of the scheduled departure, the expedition team held a grand press conference. The backdrop was *OV*, and members of the press and general public faced the team. The Discovery Channel published formal press kits which included several pages of expedition goals, promotional materials and bios of the team members. It also included several photographic slides — one of *Titanic*'s bow — so the media could use those images in promoting NBC's *Titanic Live* show as well as the Discovery Channel documentaries. The kits were given to the press and members of the public as they arrived. The expedition team had a brief meeting to explain what would happen in the press conference. The team's role was to be available for questions and interviews afterwards.

Beginning with general statements from NBC and Discovery, the conference progressed to Tom Dettweiler answering questions about the technical operations. The expedition leaders had anticipated questions about raising The Big Piece, and Tom was qualified to answer. George and PH had not arrived by the time of the conference. After Tom answered questions, the team responded to

Preparations for a New Expedition

questions from the press. The excitement of being part of this venture was evident as each expedition member smiled and spoke about the opportunity that lay before them. One member asked the audience to imagine what it would be like to see the hull piece of *Titanic* as it was raised out of the ocean. The press began to sense the emotion and magnitude of this project, and members of the general public were captivated by that image. It was a great start to the expedition.

> I stood in the back of the group, letting the important people stand up front. After a short while, I wanted to watch the conference, so I quietly slipped out to the press side. I walked around observing the general public as they watched and listened intently. Their eyes were focused on the speaker, absorbing each word just as *Titanic* enthusiasts typically do. Near the end of the conference, one member of the press had a specific question for a team member. I did not hear the question, but saw Tom Dettweiler turn to look for someone. He could not find the person needed to answer the question, and began to ask others, "Where is he?" After a moment, he returned to the microphone and asked "Has anyone seen Bill Willard?" Surprised, I raised my hand and said, "I have, a few minutes ago . . . ." Tom laughed and called me to the microphone. It was my turn to answer several questions from the crowd.
>
> Bill Willard

Following the questions and answers, the different reporters and camera teams conducted on-camera interviews with expedition members, moving from one interview to another. Groups of the general public surrounded those not involved with interviews, and everyone on the dock was engaged. As the press teams finished and departed, the expedition members thanked those who attended and returned to the ship to prepare for departure.

Our Story

> Since the departure of *Ocean Voyager* had been delayed four hours due to the late arrival of one expedition member, Dave Livingstone and I went to see the Titanic Exhibition. The exhibit had many interesting artifacts that had been collected from the debris field, but their sight made one think of those who had perished. During the tour of the exhibit, I met Charles Haas and John Eaton, notable historians on the *Titanic* story who have published books on the subject.
>
> Bill Garzke

While *OV* was preparing for departure in Boston, *Nadir* departed from the Azores, the *Abeille Supporter* sailed from the coast of France, and the *Petrel* would set out from St. John's in Canada.

Boarding the *Petrel* were many NBC technicians and several scientists and historians, including Steven Biel, author of *Down with the Old Canoe.*

> I met George Tulloch in 1996 when we both were invited to be guests on the *Charlie Rose Show* along with Peter Stone [*Titanic the Musical*]. We had the opportunity to talk in the green room before the show. In early 1998, George called and invited me to lunch. He was in Boston organizing the exhibit to open at the seaport. It was during that lunch that he raised the possibility of [me] coming on the expedition.
>
> Steven Biel

As team members arrived at the dock in St. John's, there was no ship to board as planned. There was confusion. For an unknown reason, the contracted ship was not available, and because of time, the only vessel available was an old, stripped down research vessel.

> I was in St. John's, Newfoundland, with lots of NBC people ready to board our vessel for the wreck site.

> The agent in charge came and notified us that there were contract issues and problems with the original ship booked for the trip, and we heard many stories, including that another ship was brought out of mothballs for us, and would be ready shortly.
>
> Steven Biel

The *Petrel V* was definitely in sorry shape. Engines had to be checked and readied, and electronics had to be added to comply with maritime standards. A crew had to be recruited. Two of the deck hands literally had never been to sea before. In the lounge were six chairs of different styles and a large cardboard box with paperback novels. The chairs had cost $8 each (several still had a price sticker on them) and the books were labeled "Buy all for $5." In storage areas were numerous pallets of bottled water and Gatorade. Rust spots decorated the entire deck.

> The *Petrel* finally arrived. Since we were already a few days late to the site, we were all pressed into work loading the ship – our group formed into a fire brigade to shuttle in the supplies so that we could get going.
>
> The *Petrel's* crew was great, but the ship itself didn't make a positive first impression. It probably hadn't had a chance to air out since the 1970s. I was assigned a cabin on the lowest deck with Roy Cullimore, and took the lower bunk. It definitely stirred up my claustrophobia.
>
> Steven Biel

In a matter of hours, all ships would be on a course set for *Titanic*'s wreck site.

~ 4 ~

On Our Way

July 30 - August 4, 1998

According to the schedule, *OV* was to leave Boston in the late afternoon. Word came that one final expedition member had not arrived yet. George Tulloch and PH Nargeolet were in high gear, moving from one part of the ship to another, checking in with everyone, making sure the team was ready to depart as soon as this last arrival was onboard. The entire operation was not only multi-faceted, but utilized multiple vehicles to accomplish many different goals, all nearly simultaneously.

> The month-long expedition financed by Discovery Channel, NBC & RMS Titanic was separated into two segments and had multiple operations running simultaneously during both segments. However not all operations benefited all partners. We had roughly 30 days to: 1) have a live Discovery/NBC broadcast from the site transmitted via satellite; 2) raise a multiple-ton piece of the *Titanic* hull for transport to Boston for display in an RMS Titanic exhibition; 3) raise other *Titanic* "artifacts," both large and small (including the Marconi

Transmitter) from the ocean floor for display in RMS Titanic museum exhibits worldwide; 4) photo-mosaic scan the entire wreckage and debris field, foot by foot, so that a forensic, reverse-engineering analysis could be done. This analysis would be the backbone of one of the Discovery Channel documentaries, hopefully yielding a minute-by-minute chronology of the iceberg impact and initial damage, sinking and eventual structural fatigue and break-up of the ship, the further damage and break-up as the bow and stern descended separately to the ocean floor with the ensuing debris field pattern. [Additionally, we would] 5) send scientists and engineers down to the site in deep-sea submersibles for up-close observation of the wreckage [and for use in] Discovery's documentaries; and 6) collect hours and hours of assorted footage needed to produce 2-3 Discovery Channel documentaries for future broadcast.

There were multiple ships, submersibles and ROVs at the site to accomplish the expedition goals. RMS Titanic was on IFREMER's *Nadir*. The *Nadir* also housed the manned, deep-sea submersible *Nautile* as well as the unmanned ROV *Robin*. The *Abeille Supporter* and its ROV *Hysub* were charged with the huge task of raising a multi-ton piece of *Titanic*'s hull and transporting it back to Boston. NBC production crew and talent were on the *Petrel* and I think the satellite feed and control room was on yet another boat. Discovery/Stardust production crew, scientists, naval architects and historians were on *Ocean Voyager*. The *OV* also housed two unmanned ROVs . . . *Magellan* and *Remora*."

Charlene Haislip

This was a day to meet with everyone involved with the two-hour science event scheduled to air on NBC on 16 August. There would also be television coverage when

> the large structural section from *Titanic* was brought to the surface. Sarah James and Bob McKeown of NBC were aboard and would begin to conduct interviews with some of the team. Tom Dettweiller of Nauticos had a brief meeting during the morning with the scientific crew to go over some of the items that were planned during the time on site. There was an intent to have the mini-ROV *Robin* explore some of the interior of *Titanic*. Greg Andorfer met with David Livingstone and myself to go over objectives for encounters with the stern and bow sections of the wreck. Everyone then gathered for a meeting on the general objectives and purposes of this voyage to the wreck.
>
> Bill Garzke

Meetings were typically held in the area to the rear of the bridge. When a large meeting space was needed, everyone met in the galley. Tom Dettweiler talked to landlubbers specifically about one very important point. He told the men that if they had to relieve themselves, *not* to do so over the rear fantail. The men laughed. "I am serious," he said. "If the ship rolls and throws you off balance, you will fall off the back of the ship and may never be heard from again." Again, there was laughter but also understanding that he was indeed serious. "There have been men to disappear falling over with no one around to sound the alarm."

> One of my shipmates, a Seattle photographer and writer named Paul Souders, had traveled and worked in the Kalahari Desert, the fjords of southeastern Alaska and the Siberian Taigonosk Peninsula. But the thought of bobbing around in the middle of the North Atlantic made him consider, at least for a second, the wisdom of jumping ship just before the *Ocean Voyager* left port.
>
> Susan Wels

The final expedition member arrived, and everyone prepared to get under way. Many team members lined the railing as Boston's lights diminished into a glowing hemisphere on the horizon. Each individual experienced the same sense of awe: "We are on our way to *Titanic*!"

> Last night, a few minutes after 11:00 P.M., our expedition ship, the *Ocean Voyager*, slowly pulled out of Boston Harbor, heading 950 miles due east for the *Titanic*. Landlubber that I am, I found myself thinking, as I watched the comforting, familiar city lights recede into the darkness, "Am I crazy? What am I doing here?"
>
> Susan Wels

Expedition participants were limited in number by bunks available on the ships. Beds were a critical logistical issue! But even with months of counting and coordination, *OV* actually overbooked by one [person], which caused its six-hour departure delay, major hurt feelings and someone sleeping in an "undesignated" area. Then while at sea, some people needed to change ships depending on their work. Swapping bunks was an ongoing management activity. I was lucky in having the top bunk in a cozy two-bunk cabin on *OV* throughout the trip. But my cabin mate changed several times. NBC's Sarah James was an inspiring roomie existing on a few hours' sleep and always cool, in control and gorgeous in the face of my frazzled, rapidly-aging self. Next was Jack Eaton. Clearly gender was not a consideration where sleep space was concerned. People needed to be where they needed to be! But Jack's incredible gentleman's sensitivities were strained by the abandonment of such things at sea. There was no room or time to hold doors for the ladies or help them with their cases when they were decades younger and always in a hurry. He moved on to share [a cabin] with a guy he did not feel he needed to chivalrously protect, and I got a (to be unnamed) female production assistant who only brought one pair of shoes. The smell of my new roomie's footwear was eye watering. We swiftly agreed that she would hang them outside the cabin, and [we] would frequently hear the coughing of passersby. But somehow sleep always came quickly.

Maureen Lemire

I was on *OV* when we departed from Boston. "We" means Stardust Visual and our production team, editing and camera crew, scientists and other experts, and the unmanned ROV *Magellan* and its Oceaneering crew out of Houston, Texas. *Magellan* came with a boxcar-sized control room that was securely welded to *OV's* fantail as that is where the Oceaneering crew would be "stationed" during the expedition. *Magellan* would be deployed, maneuvered and retrieved by its "crew" working out of that control boxcar. But our troubles started from the onset. *OV's* captain decided to leave port after *Magellan's* control room was securely affixed but before *OV's* non-functioning desalinator was repaired. Yep, that's right. We left port knowing we didn't have drinking water! Or, rather, only a couple of days supply. So en route to the *Titanic* site, we made contact with the NBC crew who would be departing from St. John's, Canada on the *Petrel*. We arranged for them to bring a billion bottles of water to the site!

Charlene Haislip

Quite a few of *OV's* team stayed up late, mostly due to the adrenaline and excitement of the departure. Small group conversations broke out in the galley as individuals started the process of becoming a team. By early dawn, some stood on the deck as the sun began to rise. The ship was sailing eastward, and only a few small, wispy clouds mottled the sky. Several went by the galley for a cup of coffee. There was something odd in the motion of the ship; they felt an exaggerated rocking motion.

> We're headed for the *Titanic*, the most famous shipwreck in the world, on a vessel packed with millions of dollars' worth of remote-controlled robotic cameras, fiber-optic cable, satellite equipment and video production gear.
>
> Susan Wels

> Life at sea is just like floating on a big swimming pool of water, except there is no easy way out. When the cameraman (Mark Knobil) and I accepted our mandatory sentence to spend six weeks afloat on the second RMS Titanic expedition . . . we took several mental precautions. We agreed that we would not count days, or days remaining at sea. We [would] just accept that life on board is our life, always has been and always will be . . . .
>
> Saul Rouda

Slowly, more of the team found their way to the galley. Posted on the small white board was the meal schedule and the morning breakfast menu.

Once everyone was seated, they heard an announcement about it being "Gumby suit" day. Those who were at sea for the first time curiously observed the veterans as they laughed. Another announcement called for two meetings; one before and one after lunch. Meanwhile, the two galley stewards were bringing the dishes out, greeting the team members with their heavily accented, "Good morning!"

Our Story

> Our first day at sea was 31 July and there was a lifeboat drill and demonstration of using a survival suit. This is important as survival in the sea depends on protective clothing to guard against hypothermia when you are forced to be in the water, even for a short time.
>
> Bill Garzke

> This morning, as I climbed the metal stairs to the upper shelter deck, clutching my bulky foam-filled life vest and survival suit, I thought a lot about the *Titanic* – and the fact that it's a direct result of her sinking that these drills are compulsory today. The *Titanic*'s passengers and crew were, strangely enough, supposed to have a lifeboat drill on Sunday, April 14, the day she hit the iceberg, but Captain Smith decided to forego it. It's questionable how much difference it would have made, since there weren't enough lifeboats anyway.
>
> Susan Wels

In each cabin was a small carry bag for each occupant, much like a thin sleeping bag. The team members carried their bags to the designated deck, where they were to completely open the suit, climb into and seal it — all within two minutes.

> The *Ocean Voyager*, by contrast, has more lifeboats than we need – six inflatable rafts that can each hold 25 passengers and crew. And as I awkwardly wriggle on the floor into my neoprene survival suit – a full-length hooded straightjacket that makes me look like a fluorescent orange Gumby – I feel confident I can survive whatever the ocean dishes out, so long as I have unlimited time and at least three people to help me squeeze into this unfathomably complicated gear.
>
> Susan Wels

The suit covered the entire body except the face. Attached to it was a whistle, to blow if you were in the water needing to get the attention of a search party. Several at a time, members took their turn under the watchful eye of *OV's* Captain. All of the team members passed the test. Jack Eaton, a senior member of the team, had the most difficulty. Fortunately someone continued to distract the Captain while others "assisted" Jack. When the Captain called "Time!" and turned around, Orange Gumby Jack stood there sporting a huge grin and outstretched arms! After these tests, the suits were rolled again and put back in the carry bags. Little did anyone know that would not be the last time they would see these suits.

> Saturday, 1 August began with *OV* now well out to sea. Greg Andorfer called a meeting today to discuss the bow encounter of *Magellan 725* with *Titanic,* and in attendance were Charlene [Haislip], Bob [Sitrick], Gary Hines, Bill Willard, Dave Livingstone and I. This was important for the proper perspective for opening the documentary, *Titanic: Answers from the Abyss. Magellan* would start forward then work its way aft. It was decided to leave the survey of the mid-ships expansion joint for the submersible *Nautile* and *Robin* at a later time during the expedition.
>
> Bill Garzke

The 2,000-ton *Ocean Voyager* is definitely no greyhound. Today, we're chugging along at an average speed of about 9 knots – far less than half the speed of the *Titanic* and considerably slower than a waterskiing boat. That may be a blessing – especially because, even at this speed, the *Ocean Voyager* is seesawing wildly from side to side in relatively tranquil seas. Needless to say, many of us have opted to pass on the lunch menu of goulash and lemon meringue pie.

Instead, we're scattered out on the decks seeking fresh air and sunshine, which has thankfully arrived.

Susan Wels

After lunch, the work continued. The rocking motion of the ship had increased. Not only was the ship rocking forward and aft with the direction of the ship, but there was a side-to-side motion as well. Several people discussed this issue and came to the conclusion that the extra weight welded to the top deck caused the ship's balance to be exaggerated. The majority of the team opened seasickness pills that first day. The experienced crew of *OV* adjusted the ballast and made even more adjustments on the third day, so that the rocking motion was closer to normal.

> The ocean, thankfully, is gentler today, and most of us have recovered from the effects of yesterday's lurchings. Bob Sitrick, Discovery Channel's technical operations director, has unrolled his beach towel and is catching some rays on the top deck. But not for too long. As Jimmy James, the ship's grey-bearded safety officer, tells me, there's no such thing as a Sunday out at sea.
>
> That's definitely true on the *Ocean Voyager*. Inside the bridge, the meetings go on, hour after hour, with the expedition's naval architect and marine engineer, David Livingstone and Bill Garzke. The two experts huddle endlessly with the television team over a detailed model of the *Titanic*'s wreck and debris field, debating strategies for surveying the ship's mangled stern.

Susan Wels

One of the meetings called was to discuss an entirely new problem.

> A serious problem arose during the afternoon when it was discovered that the [Dynamic] Positioning System could not operate when *Magellan* was deployed. Technicians from Oceaneering began working on this dilemma to solve the problem.
>
> Bill Garzke

This issue was significant. Imagine a ship on the surface, bobbing up and down with the motion of the waves as it drifts with the surface currents. Beneath that ship, attached on a 4,500-meter tether, is a tremendous weight. The upper ship must stay relative at one spot. The Dynamic Positioning System is comparable to an on-location cruise control. When the system detected movement away from a designated spot, it would activate the ship's thrusters and relocate the ship back to the intended position. Oceaneering, the team that operated and maintained the *Magellan* ROV, would not deploy unless this system was fully operational. If the ship's propellers were to sever the tether of the ROV it would be catastrophic to the expedition.

Early on the morning of August 2, work was under way. After breakfast, the television crews were huddled with Bill Garzke and David Livingstone. Someone affectionately gave them the nickname of "Garzke and Hutch" and the moniker caught on quickly with the Americans. "Who is Hutch?" David Livingstone, from Belfast, Ireland would ask.

> On Sunday, 2 August, discussion continued on the bow and stern encounters. Since Dave Livingstone made two dives to the wreck in 1996 with the submersible *Nautile*, [it was decided that] he should be the one to lead the discussion on important points of the stern wreckage.
>
> Bill Garzke

Our Story

The rest of the team — those who were not needed in the meetings — found ways to enjoy the trip. The bridge had several sets of binoculars, and it was thrilling to scout for whales, looking for the water spout as they surfaced. Crew members were always eager to point them out to those searching. Several members of the team would read or bask in the sun.

> On an afternoon like this, the North Atlantic is a fine place to be. Sitting in the sun outside the bridge, I'm almost hypnotized by the rolling sapphire sea and the cottony clouds hugging the horizon.
> What's most amazing to me is the wildlife out here in the middle of the ocean. Herring gulls, cormorants, kittiwakes and other seabirds skim the surface of the sea, diving for food hundreds of miles from any land. Though I haven't seen whales today, I did spot the dark dorsal fin of a shark just a few ominous feet off the starboard side.
>
> Susan Wels

NBC requested small interview clips with team members who had been promoted in the earlier press release. These clips, consisting of a basic name, role in the expedition and the answers to two or three questions, would be shared within their network, providing local news affiliates throughout the nation an opportunity to request interviews with one or more of the team. The interviews could be live via the NBC satellite, or taped for later viewing.

At the same time, initial segments for the planned programs were scheduled for filming.

> In the evening Dave Livingstone and I made a voice recording illustrating important features of the bow encounter that would be used in the opening of a planned television documentary for RMS Titanic, Inc.
>
> Bill Garzke

Monday, August 3, 1998

This day was overcast, and many of the team spent moments in the galley in various conversations. Word was received that *Nadir* had arrived at the location of The Big Piece.

> This afternoon, George and the crew of the expedition ship *Nadir* did come back to the patch of ocean above the hull section's position, and they successfully located The Big Piece once again. This time, they're certain that they'll raise it. If they succeed in a few days, it will be the first time that the *Titanic*'s portholes will be above the waves since 1912 – in a small way, a resurrection. The good news from the *Nadir* was a bright spot in an endless day of grey skies, grey rain and grey water. We are making our way, slowly, to the *Titanic* site, and the *Ocean Voyager's* technical and television crews are completing their last plans and preparations.
>
> Susan Wels

The Stardust team members were working hard to prepare for the shooting schedule.

> I decided to make a sketch of how *Titanic* looked when Second Officer Lightoller swam to the lookout's position on the foremast to escape from being pulled down with the sinking ship. This sketch revealed that the trim of the ship was greater than 15 degrees! This meant that many compartments in the forward section of the ship were flooded or about to be flooded. It also meant that extreme stresses were being placed on the amidships area of the ship. This would account for the large stresses in the main deck and keel that were found during the 1996 finite element analysis. Dave Livingstone, after reviewing the sketch, felt that a more detailed analysis should be made. Greg Andorfer

concurred and Gibbs & Cox, Inc. were called upon to do that analysis.

Bill Garzke

In the evening, moods were light. The team knew they should arrive on site during the next afternoon. People began to muster in the galley where there was a TV and a VCR. A team member named Jeff, who was part of the satellite dish network, had rushed to the store just after the press conference, and purchased a collection of VHS tapes for the journey. The tapes were the first episodes of the animated sitcom *South Park*. Many of the team had not watched the show; there was loud laughter at the on-screen antics from those who followed the show, and even louder from members who were seeing it for the first time. Everyone sat through two episodes, otping to save the others for another time.

Tuesday August 4, 1998

At 1:40 A.M. this morning, there was no sky, no waterline, no stars or moon, as if a black curtain had been dropped on the Atlantic. With my land-dweller's eyes, I couldn't make out anything on the invisible horizon. But out on the port bridge, our navigating officer, Captain Tony Foster, and safety officer Jim James had their binoculars focused on the impossibly faint lights of a small vessel, the *Verna and Gean*. The 100-foot boat had sailed out of St. John's, Newfoundland, to bring the *Ocean Voyager* a differential global positioning system (DGPS) – navigation gear that would help our ship remain in a fixed position over *Titanic*. Without the system, it could be dangerous to launch the ROVs that are tethered to the *Ocean Voyager*. If the ship drifts beyond a certain tolerance, a crew member explained, it could drag an ROV over the wreck like a dog through a briar batch.

> So several of us came out here in the early hours to watch the mid-ocean rendezvous, set for 2:00 A.M. As we stood on the bridge deck, heat lightning flashed over the rough water, and sea birds – stormy petrels, called wave dancers by sailors – flew like scraps of paper in the rising wind.
>
> By 2:13 A.M., the *Verna and Gean* had drawn close by. Our Zodiac driver, Julien Nargeolet, readied his motorized craft to make the pickup, and the expedition's webmaster, Matt Tulloch, gamely strapped on his radio transponder and safety vest to ride along. At 2:35 A.M., they roared off into the dark, accompanied by the boatswain, Les Murdoch. Ten minutes later, the Zodiac reappeared, smashing its way back through high black waves to the *Ocean Voyager*, loaded with crates of navigating gear. Matt, at least, looked happy to be back as he climbed out onto the deck: "That," he said, as the sea crashed over the fantail, "was a scarier trip than I expected.
>
> Susan Wels

In the morning, many of the team had not known about the delivery during the night. For most, it was back to the busy routine of preparing for the productions.

> I continued my analysis of the large trim on Tuesday using data that Dave Livingstone had brought from Harland and Wolff. Lightoller's testimony at the hearings indicated that the waterline before the plunge was tangent to the underside of the Crow's Nest. He also noted that the waterline came to the bridge level. At this point the ship was doomed to sink quickly. However, Lightoller was not in a position to see the hull failure that was observed by several other survivors, particularly Seaman Bulley, who testified that he could see into the Reciprocating Engine Room from

> the lifeboat he was manning. After discussing my findings with David Livingstone and Greg Andorfer, they were pleased to find that my rough calculations were a good indication that the ship took an unusually large trim that lifted the stern clear of the water surface.
>
> Bill Garzke

In France, the *Abeille Supporter* was finally underway, headed for the wreck site.

> Our late departure from Bayonne, caused by the French government's last-minute call-up of the *Abeille*, caused general concern about whether we would arrive at the wreck site in time for the scheduled live television broadcast. Speaking of broadcasts, Jack and I felt a bit news-deprived on board, even trying to find an English-language newspaper when *Abeille* docked at San Miguel in the Azores. Another day there was great excitement upon spotting and recovering a drifting red lifejacket in mid-ocean. The vessel's name had been painted out, but the jacket had been manufactured in Newcastle, so we knew it was not from *Titanic*.
>
> I found myself in a claustrophobic bunk, my face just a few inches below the ceiling in *Abeille's* cabin 101 on the main deck. At 3:30 A.M. on the first night at sea, a grinding sound jolted me awake; at first I thought we had run aground. It was only the ship's bow thruster (apparently directly below our cabin) maneuvering the ship as we warped into Bilbao, Spain for fuel bunkering. I deployed my earplugs thereafter.
>
> We spent hours at the sun-drenched picnic table on the upper deck, the only place (aside from our cabins) where one could spend the day, watching dolphins, flying fish, waterspouts and the ever-changing sea while engaging in long, wonderful conversations with our teammates. But there were other diversions, including

the use of a shipboard shower that often resembled a lake, and several shipboard emergency drills, featuring the donning of the dreaded immersion suits.

Charles A. Haas

With news that *Abeille* was on its way, the final piece of the puzzle was aimed for *Titanic*'s site. Predicting ETAs gave planners a starting point for the master planning schedule. And plan they did.

~ 5 ~

Arriving at the Site

Tuesday, August 4, 1998

A PERSON CAN TRULY UNDERSTAND the concept of "insignificant" when standing on the highest deck of a ship at sea, panning 360 degrees to view nothing but ocean, ocean, and more ocean. There are no landmarks to identify location, no signs or billboards either. But storms are visible in the distance as they produce rain. The sun's rays paint a majestic canvas in myriad patterns. The tide rolls for miles and miles, the sun glistening and sparkling in an unending random array of flashes.

As *Ocean Voyager* traveled for a final day, the expedition team was to arrive at one significant location in the middle of that vast ocean, where *Titanic* sank beneath the waves 86 years earlier. Team members joined one another on the top deck, wanting to see the site from the best vantage point. A change in the tenor of the ship's engines let them know the ship had arrived at its destination. Cameras clicked as the ship's motion slowed to a gentle roll. Historians, video specialists and technical crew stood side by side in respectful reverence.

Arriving at the *Titanic* site is always a very emotional experience. We have all read the stories and seen the films, and it is very touching to arrive at that spot in the ocean and think about what happened there....

Bob Sitrick

Quietly, each person moved downward to the bow of the ship where, in silence, they tossed roses into those hallowed waters. The silence continued as some experienced deep emotion. Many eyes were filled with tears for all those who perished in that tragic sinking.

It's nearly 4:00 P.M., and we've finally reached ground zero, the *Titanic*. The sea is lead grey, except for the broken crests of waves. We know that miles beneath us is the wreck of the great ship, and we drop flowers in the waves.

Susan Wels

After a long period of reflection and remembrance, the team departed in small groups as everyone had work to do. Work returned to normal.

That afternoon, Paul Mathias came aboard *OV* from *Nadir* with PH Nargeolet and George Tulloch. A discussion was held [about] the importance of keeping the ship on station during the two-hour *Titanic Live* documentary scheduled for August 16 on the NBC network. Paul was to be aboard *Nautile* that night and would maneuver the mini-ROV *Robin* [from] inside the reciprocating engine room in the stern section of the wreck.

Bill Garzke

Later in the evening, silent observers assembled one or two at a time. Some stood back from the deck's edge, others leaned over, arms propped on the ship's railing, but all were gazing into the dark waters highlighted only by the moon's reflections blinking nonstop on thousands of small wave peaks. Those who knew the *Titanic* story well stood beside those who were just beginning to learn it. Someone announced, "Ship's time is now 11:35."

> It is one thing in the daytime but even more moving during quiet moments at night – gazing out into the ocean and imagining how terrifying it must have been that fateful night. I have now been to the site three separate times, and on each occasion we held a memorial service where we cast flowers into the sea and observed a moment of silence.
>
> Bob Sitrick

> I openly wept quietly. Tears were streaming down my face . . . . I could see them in the water. It was so quiet, I almost heard them calling for help.
>
> Anonymous

The calm at the site was short-lived, however. A storm hit, and hit quickly. The next morning, damage caused by the storm was evident. The production teams renewed their efforts now that the ship was no longer traveling but on site. Concern rose over *OV's* capability to stay positioned during *Magellan* dives. The remnants of the storm still rocked the ship. As the seas calmed, the new positioning system was in place, and all eyes would be on the test — to see if *OV* could maintain its position.

> During the overnight period of August 4 and 5, *Ocean Voyager* was buffeted by a storm that caused damage to the ROV *Remora*. It broke free of its lashings and cracks were sustained in two of its struts.
>
> Bill Garzke

One of the Oceaneering crew was called up to secure the *Remora*. He tied a rope around his waist and jumped into water that was often waist deep as the storm surge rushed in and out over the fantail. Several of the Oceaneering team were above, holding the other end of the rope.

Bill Willard

Things haven't worked out quite so well, though, for the *Ocean Voyager's* ROV technicians. Last night, in a 35-knot gale, the standby ROV, *Remora*, crashed to the deck and broke two of her struts. She's repairable – but all things considered, it was a pretty gloomy morning.

Susan Wels

Oceaneering (who was in charge of ROV operations) became concerned with the ship handling and with the bow thruster as it could not hold the ship in position. Attempts to hold *Ocean Voyager* within a 100-meter circle that afternoon were unsuccessful. Oceaneering would not commit its equipment until they could prove that better positioning could be maintained.

Bill Garzke

It looks like the new differential global positioning system is a dud.

Susan Wels

Zodiacs were boats with sides inflated by a pressurized gas. Stretched between the sides was flooring, and these boats were powered by an outboard motor. Everyone wanted to experience a Zodiac ride at least once. Zodiacs were used at the site to transport personnel and equipment from ship to ship. Imagine four-wheeling on the open ocean with sea spray hitting you as you rise and drop from each wave. The main driver, Julien Nargeolet, had mastered the art of accelerating into the wave in order to achieve a greater launch effect.

> Even at the end of the 20th Century, technology's no match for the ocean and the weather. This morning, around 9:00 A.M., Julien Nargeolet and Discovery Channel's Bob Sitrick set off in a Zodiac from the *Ocean Voyager*, headed for the French research ship *Nadir*. They almost didn't make it. Halfway between the vessels, the Zodiac's engines died, and a heavy fog blew in. They were drifting blind, unable to gauge direction, and the boat was taking water. Fortunately, the fog suddenly lifted, their flooded engines started, and they made it the half-mile across the water to Nadir. As commutes go, it was a little hair-raising, but their luck held out.
>
> Susan Wels

At this time, there were two main operations concurrently under way. Operations on *Nadir* were focused on attaching the lift bags to The Big Piece and preparing to take it aboard the *Abeille*, but the *Abeille* was behind schedule. A very late departure from France put the *Abeille* still several days away, leaving little time to spare for the targeted date of raising The Big Piece. This delay would become a greater problem if the weather was uncooperative and would cause additional delays. Meanwhile on *OV*, *Magellan* had still not launched, and two television production teams were stressed. There was no *Titanic* footage for a show just over a week away. Both Stardust and NBC shifted to capturing footage of the forensics team at the *Titanic* site. Waiting for *Magellan* to launch, a high uneasiness hovered over the production specialists.

Our leader, George, was quite adamant about following the rules. One rule was that everything from [*Titanic*] had to be quarantined or at least locked up so that recovered objects would not wander away. Early in the voyage the ROV gang, complete with an ex-marine leader who just oozed experience and a cool head, asked if they could have an insignificant piece of *Titanic* as a souvenir, like a piece of pipe they could divide up. I could understand their desire to have a keepsake from this glorious endeavor. They were refused and told to give up that desire. No souvenirs. Even rust (rusticles) on the deck were religiously washed overboard. It was hard to obtain any swag with George watching our every move when *Titanic* objects were being sorted, preserved and locked up.

Saul Rouda

Later in the day, Susan Wels moved to *Nadir* to cover the preparations for The Big Piece. The Zodiac ride she witnessed earlier in the day would be forefront in her mind when it was her turn to ride in the oceanic ATV.

> As we're all learning, you can't take anything for granted on the ocean. Which is why, this evening, I'm more than a little nervous as I strap on a life vest and hop into the Zodiac at sunset. I'm moving over to the *Nadir*, in the middle of the North Atlantic, on a rubber dinghy piled quite a bit too high, it seems to me, with heavy packing boxes, plastic crates and all my worldly possessions — a duffel and backpack filled with clothes and books and a laptop computer bundled, overcautiously, inside a raincoat and a plastic garbage bag.
>
> As the Zodiac pulls out and we smack into one giant swell after another, I learn another truth about the ocean: The gentle ripples seen from the top deck are

more like Himalayas when you experience them up close on the water. But our Zodiac driver, Julien, is unfazed, and despite the rollercoaster dips and heaves, he pulls up to the *Nadir's* fantail without flipping me and my gear into the water. Another day, another little triumph.

Susan Wels

Thursday, August 6, 1998

Two days later, a mechanical problem in the *OV*'s thruster system was identified. Devices that acted like scuppers to filter the fuel line were clogged. Those devices were cleaned and returned, followed by short tests. Cautioned hope broke out — it seemed to work.

> On Thursday, August 6, decisions had to be made on the ship handling. The regular ship Captain was replaced by Eric Pailles, [the executive officer on the ship, also a licensed captain] who would try to demonstrate that he could keep *Ocean Voyager* within the requirements desired by Oceaneering. There was concern as *Magellan 725* had not been sent down to the *Titanic* wreck. The problem with the ship positioning was discussed. It was [later] announced that the bow thruster was working and there was a good probability that production of *Titanic Live* would proceed. David Livingstone went up to the bridge to watch the ship maneuvering with the bow thruster. The new man behind the controls was able to keep the ship within a 100-meter circle.

Bill Garzke

[August 6] is George Tulloch's birthday. Our expedition leader has just turned 54, and he's been celebrating with a good cigar and a glass of Beaujolais. It's a fine day, too. The sea is azure, the sun is shining in a cloudless, pale blue sky, and the *Nadir* is poised above the successfully relocated Big Piece. Aboard the *Ocean Voyager*, the *Magellan* engineers are putting the last touches on their ROV before they launch it for a deep-water test. The crew has been working long days and nights checking the cameras, interfaces and fiber-optic cables. "Now," says engineer George Brotchi, "I just want to get on bottom and show everybody that it works."

As a final step, they tie a white mesh bag filled with styrofoam cups onto the underwater system's metal frame. When the equipment reaches bottom, 2.5 miles down, the water pressure – 6,000 pounds per square inch – will squash the cups into deep-ocean souvenirs of flattened foam. George Brotchi has personally collected about 200 of them, records of his most memorable dives.

Susan Wels

The day's earlier hopes that the ship could successfully stay on station and allow for a successful deployment of *Magellan* were soon dashed.

Enter another licensed captain on the ship, Tony Foster. It was Tony's turn. He would attempt to keep *OV* on station.

Bill Willard

Different people attempted the *OV* manual positioning. They let one crew member try, and Julien Nargeolet was also doing the positioning. I remember the *OV's* Captain was happy to have Julien as he was used to playing computer games and because of that, he was good to maintain the *OV* in a stable position.

PH Nargeolet

As the day wears on, however, the *Magellan* engineering team's still waiting. The ship's position must be constant while the ROV is on the bottom, and so far, the *Ocean Voyager* has been moving around too much to launch.

By 10:45 P.M., there's not much left for the frustrated *Ocean Voyager* passengers and crew to do but sit down and share some birthday cake with George Tulloch, who has zipped over from the *Nadir* for a visit. An NBC *Dateline* producer, Joe Ferullo, passes around plates of layer cake with a caution: "Have some," he says, "but the Big Piece has to be for George."

Susan Wels

Saturday, August 7, 1988

The next day dawned with encouragement for those working on the positioning system. A new test passed Oceaneering's careful eye. There was improvement with the ship's handling, and the *Magellan* was set to launch at 10:30 A.M. for a three-hour descent to the ocean floor. The heads of the production teams immediately began to negotiate what clips were needed first. *OV's* top deck was jammed with observers as the large aft crane moved into position then lifted the *Magellan* ROV off its resting place. The tether system began to release the tether, more than 2.5 miles of it, so that *Magellan* could maneuver. As the crane pivoted, the yellow behemoth slowly moved off the stern. Trained technicians were ready to act in case of complications, but there were none. Cameras clicked away, and cheers broke out among the observers. A production unit, with cameraman, soundman and scene director were filming from different locations as they captured the first launch. *Magellan* was lowered into the water, then on command from the Oceaneering team, began to disappear into the North Atlantic. Once it arrived at the debris field, it began to transmit images to *OV*.

With images coming from the debris field, PH Nargeolet assembled a meeting with Dave Livingstone, Greg Andorfer and me. Images streaming in from *Magellan 725* were extremely good, and all of us were able to identify important items. We were able to locate four boilers from Boiler Room No. 1 and one of the low-pressure cylinders from one of the large reciprocating engines in the main engine room. Many of the items strewn around these items came from spaces over Boiler Room No. 1. The debris field provides some evidence as to what state *Titanic* was in when it arrived on the seabed.

Bill Garzke

Through its miles of fiber-optic cable, the ROV sent up high-resolution video images of doors, pots, crockery, furniture, boilers and other items that spilled from the *Titanic* as she sank.

Susan Wels

Many team members relocated to the monitor room after an early lunch. They watched as *Magellan* located an object and went through different functions. On some items, the cameras would zoom in, on others zoom out. The ROV would move around to get a complete look at them. All of these motions were coordinated with the production lead to show detail and creativity.

At the same time, those in the monitor room were identifying each known piece and discussing the items that were not as quickly recognized. A person watching from the rear of the room would see some of the observers tilt their heads one way then to the opposite side to try to identify a mystery piece from different angles. Someone else would turn pages in a resource book in order to identify the pieces. Layers of light sediment or a coating of rust made identification a difficult task, but after several days of inactivity, each person involved was hyped with energy and willing to help. Before long, someone commented, "We've been at this six hours!" Everyone decided to do dinner in shifts so the work could continue. Most ate quickly and returned to the monitor room.

Slowly, the group monitoring the images from below began to leave for the evening as the announcement came that *Magellan* would be returning to *OV*. The first dive went flawlessly. While still in the room, one member of the group received exciting news.

> At 2200 (10:00 P.M.) Dave Livingstone was informed that he would board *Nautile* on August 9.
>
> Bill Garzke

The task set upon David for his dive was to ascertain the actual location on *Titanic*'s hull where The Big Piece belonged. The previous estimated location now caused doubts in some people's minds, and the production team wanted conclusive evidence of "where The Piece was located" for the live show scheduled to air in about a week.

The *Abeille* had finally arrived and was with *Nadir* at the location of The Big Piece. *Abeille*'s silhouette was simple to identify with the towering A-frame on the stern of a long work area, all painted in a bright yellow with Carolina blue accents. Those on board *Abeille* were glad to be at the site, and immediately the Zodiacs added a third ship to their routes.

> Once Zodiac service had been established between the three ships, we enjoyed visits from Günter Bäbler, Stephen Biel, Claes-Göran Wetterholm, Bill Garzke and David Livingstone.
>
> Charles A. Haas

In the evening, a smaller group assembled on *OV*'s upper deck as the *Magellan* returned home for the evening, its whitish glow growing larger under the blue sea as it neared the surface.

Now, a few minutes before midnight, the *Magellan* technical crew from Oceaneering is out on the fantail recovering the cable and waiting for the ROV to rise. Under a full moon, a blue light suddenly appears under the water, and soon Magellan's yellow body breaks the surface and rises high above the fantail on a crane.

Susan Wels

A key gang was the ROV operators and engineers. The pilots were a sweet bit of youthful crazy. They loved remote controlling the autonomous subs, a fancy video game. A mentality but without any guilt, perhaps like those drone operators who [attack] from thousands of miles away, making their deeds quite abstract.

Saul Rouda

Two of the Oceaneering guys picked up a quick nickname. Frank McKinney and Stan Jandura shared a room, and so they became "FranknStan."

Bill Willard

On *Nadir*, Paul Matthias was preparing for his dive in *Nautile*.

Aboard the *Nadir* today, imaging expert Paul Matthias, president of Polaris Imaging, has spent much of his time assembling a groundbreaking new optical system for collecting virtual 3D images of the *Titanic*. The system is made up of two high-resolution still cameras – Gabriel and Lauren, named after Paul's children – which he'll mount on the submersible *Nautile*, along with two strobe lights and two laptop computers. Tomorrow, if all goes well, he'll dive to the *Titanic* and test the new imaging technology.

"What we're doing," Paul says, "is pretty ambitious." By gathering thousands of images of the top and sides of the ship, as well as the debris field, he'll create a precisely mapped, computerized photomosaic of the wreck. The images will have three-dimensional depth, and their resolution will be so high that the smallest details will be fully visible. "We'll have pictures so clear," he adds, "that we could make them 100 feet high and you still wouldn't be able to see the pixels."

If the system works as planned, it will represent another breakthrough in *Titanic* imaging.

Susan Wels

SUNDAY, AUGUST 8, 1998

With the previous day's success, everyone was up early on August 8, looking forward to the accomplishments of the day. *Nautile* would dive with Paul Matthias onboard, with his new optical system in place. *Magellan* would launch again; the day's objective was to film the forward peak, then to reconnoiter the bow.

> This afternoon, our neighborhood traffic suddenly doubled in the North Atlantic. A container ship appeared on the horizon–the first passing vessel we've seen–and our third expedition ship, the *Abeille Supporter*, finally arrived from Bayonne, France. With her yellow A-frame on the fantail, the brawny *Abeille* will be standing by to receive The Big Piece, if it makes it to the surface. On board are *Titanic* historians John Eaton and Charles Haas, as well as the ROV *Hysub* and more members of the expedition team.
>
> Susan Wels

Nautile's launch was delayed due to technical issues.

On the *Nadir*, the test dive Paul Matthias planned has been delayed due to problems with *Nautile's* electric cables. At 3:40 P.M., the sub is finally launched – but in just two hours, *Nautile* is once again back on board *Nadir*, having dived only halfway to the bottom before the crew called the experiment quits.

Susan Wels

Paul climbs out of the *Nautile*, a little frustrated but clear about the work he has ahead of him. "We now know where the weak points are in our cabling system," he says. "For some reason, water was able to penetrate the off-the-shelf equipment we installed, though our laptops and strobes worked like a charm. Fortunately, we've got time to get everything functioning before our imaging project gets fully underway on August 20."

Susan Wels

But *OV* had an issue of its own — the positioning system problem returned.

Frustration seemed to be the theme of the day on the *Ocean Voyager*, too. The vessel's bow thruster blew out again, making it difficult for the ship to hold position. As a result, *Magellan* couldn't be launched until late into the night – and then the ROV had its mission suddenly aborted when its cameras failed.

Susan Wels

[There was] mild boredom, or just a passive acceptance of life aboard ship . . . but for sure, life got more interesting especially when word came about the Perfect Storm in the Atlantic with three hurricanes "a brewin' ". . . . Something like the storyline in the then-current best seller *The Perfect Storm*. Word was that no copies of the book were allowed onboard.

Saul Rouda

There was a production/dive schedule that had been agreed upon by all parties prior to the start of the expedition. But needless to say, things change and complications arise once you are out at sea. For instance, *Titanic's* bow and stern were a half-mile apart on the ocean floor. The plan was to have *Nautile* down at either the bow or the stern with *Magellan* simultaneously down at the location *Nautile* was not. But depending on the ocean current speed and direction, the "swag" in *Magellan's* cable might wander too close to *Nautile's* location, endangering the lives of those aboard *Nautile*, even with the half-mile separation. Needless to say, discussions, arguments, fights and dirty tricks ensued each and every day as people jockeyed for the dive time that they knew they needed.

Discovery Executive Producer Maureen Lemire had been designated as the overall Expedition Manager. Hers was supposed to be the final word. But I can still see the look on her face the first time her decision was ignored. On that particular day, extremely strong sub-surface currents made a double dive impossible, so either *Nautile* or *Magellan* could descend, but not both. Maureen decided very late that evening that *Magellan's* production footage was more important than *Nautile's* artifact retrieval. *Nautile* was therefore told to stay put on top while *Magellan* was readied to deploy early the next morning from the *OV*.

[This was] all fine and dandy except the RMS Titanic people totally disregarded this plan. *Nadir's* crew busted their butts to get *Nautile* checked, serviced and back in the water in record speed. So by the time *Magellan* started to deploy the next morning, *Nautile* was already on the ocean floor. *Magellan* had to stay put, the production footage was on hold, and there was not a damned thing Expedition Leader Maureen Lemire could do about it.

Charlene Haislip

Arriving at the Site

In a discussion with Susan Wels, Gary Hines calmed the frustrations a bit.

> You just have to take it as it comes out here on the ocean. You're dealing with the weather, with metal and the sea. Plans and schedules don't mean too much out here – but then again, that's all part of the story.
>
> Gary Hines

To express the situation in the terms of an NBC specialist — "Here we are right at a week before one of the biggest TV specials — and it's a live special — and we have very little banked footage to draw from. We have no main shots and only a few days to put this thing together." It was agreed that *Nautile* and *Robin* could get the needed footage in the event *Magellan* was still unable to deploy, providing the weather cooperated.

With all the setbacks so far during this expedition, team members laughed and said, "Well, everything that could go wrong has gone wrong — at least everything from here on out will be good news."

> Then again, there's that hurricane we hear is forming in the south . . . .
>
> Susan Wels

~ 6 ~

Recovery of The Big Piece

August 9, 1998

*N*ADIR WAS SWARMING with crew, film teams and historians. During the previous days, *Nautile* had attached the five lift bags that were estimated to be enough to lift The Big Piece. At one time, Pierre Valdy called for up to seven lift bags, but he lowered that number. He did not want the Piece to accelerate as it rose to the surface.

David Livingstone and Bill Garzke ("Garzke and Hutch") spent the pre-dawn hours studying the photos of The Big Piece to ascertain its original location on the ship's hull. An initial opinion from 1996 had become uncertain.

> Dave Livingstone and I were up early Sunday morning, August 9, to discuss just where "The Big Piece" was located in the ship. We decided, based on the plans we examined, that it was from C-Deck on the starboard side in way of Cabins 86 and 88. This information was passed

to Charles Haas and Jack Eaton to see if they could identify what passengers were in this area of the ship.

Bill Garzke

This was also a day David Livingstone had eagerly anticipated. He departed *OV* by Zodiac at 7:00 A.M. for *Nadir* and his diving assignment in *Nautile*. Crews on *OV* were already active. On this day, *Magellan* would also dive to reconnoiter the bow. This initial study of the bow normally provided much information: What elements of decay are evident since the last dive? What parts of the bow have broken or fallen off since the last expedition? The images captured often revealed new, larger rusticle patterns, much to the interest of Dr. Roy Cullimore and his assistant Lori Johnston. The images also gave the production teams ideas for what they would capture on film and feature in their documentaries.

The excitement on *OV* quickly crashed, however, as once again, technical difficulties interfered.

Aboard *OV*, *Magellan* was deployed to make a survey of the bow wreck of *Titanic*. Soon problems arose with the compressor and the winch that deployed the cable. *OV* sustained a power failure which caused problems with the generator providing power to the ROV. Technicians were busy trying to rectify the problems. The electrical load had exceeded the capacity of the ship service generators. The electrical demand had to be lowered.

Bill Garzke

Those who were on *OV* were dividing their attention between the electrical issues and the news from *Nadir* — was The Big Piece on its way to the surface? With the power failure, no news could be received. On *Nadir*, eyes were on the monitors or the ocean.

> Now it's a waiting game. Five lift bags filled with lighter-than-water diesel fuel have already been attached to The Big Piece – enough, theoretically, to lift it. So far, the huge section of the *Titanic*'s hull plating has shifted into a more vertical position, but it still seems to be stuck 11,000 feet down on the bottom. "It may come up today, it may come up tomorrow," shrugs Pierre Valdy, the expedition's mission chief. In case it does appear this afternoon, a dozen of us have staked out positions on the top deck of the *Nadir*, where we have an unobstructed 360-degree view of the Atlantic. It's a calm, almost windless day, and we're all keeping our eyes out for an orange lift bag in the miles and miles of open sea. After an hour and a half, though, most of us give up and head indoors. Pierre Valdy figures that suction is keeping The Big Piece on the bottom, and he's decided to add a sixth lift bag in the morning.
>
> Susan Wels

Just after lunch, the *OV* team received word that the final ship contracted for this expedition, the *Petrel*, was approaching. The *Petrel* carried several historians and scientists, but mostly the bulk of the NBC crew that would prepare and produce the *Titanic Live* show. From the top deck on *OV*, observers saw the bobbing silhouette in the distance. Officially known as the *Petrel V*, this vessel was quite narrow, with the center console toward the stern. The bow would bob up and down, while amidships would remain relatively stable. As *Petrel* neared, *OV* greeted her with a horn salute while the team aboard *Petrel* responded with waves. The team for the first phase of the expedition was finally complete.

> We stood on the deck of the *Petrel* as we approached the site. When we neared the flotilla in front of us, we knew "we're here."

> Arriving at the location where the *Titanic* sank – it was a feeling I didn't anticipate. There was an unexpected eeriness about it – and a quiet contemplativeness. On the other hand, it looked no different from where we were 15 minutes before – so there was a strong competing feeling, something like "So this is it?"
>
> Steven Biel

We arrived at the site before NBC did. We kept an anxious lookout and were thrilled to see them on the horizon just as the sun was setting one evening. Elation turned into howls of laughter as we got a closer look at their ship. You see, they insisted on chartering their own boat themselves, apparently not trusting that we (Stardust Visual) would book something "good enough." NBC, after all, had all the "talent" on their boat. Talent that required civilized accommodations. In this case, looks were not deceiving. The *Petrel* looked like some rickety pirate ship from the 1500's. It had that severe "sloop" mid-ship, and one could not help but hallucinate that the whole thing was made of wood rather than metal. As it got within radio distance, we were informed that they were very anxious to board the *OV* the minute they arrived so we could get our water.

We radioed back our thank you's, but said they'd have to wait until the next morning to board our ship. You see, we were deploying *Magellan* at that time. "Deploy" meant a slow, steady, controlled drop via tethered cable to the ocean bottom some 2.5 miles below. It was very dangerous to have any other boat or ROV or anything at all close to our ship while the tethered *Magellan* was deployed with its miles-long cable waving around behind it. Boats and ROVs could get tangled up, lines could get cut, and that multi-million-dollar piece of equipment could get lost to the deep forever.

The NBC folks were extremely disappointed at the news that they could not board *OV* the moment they arrived at the site. You could hear what sounded like distress in their voices and we found this very curious. We thought, "How very nice of them to be so concerned about our water situation." Well, as it turned out, our drinking water was nowhere on their list of worries. You see, the toilets on that fabulous boat they had chartered were not working. In fact, they had ALL stopped working shortly after leaving St. John's. And yes, that included toilets for their "talent" as well. The stench was so overwhelming that everyone stayed up top, on deck at all times. And here we thought everyone was on deck waving and calling out to us just because they were so happy to see us!

Charlene Haislip

I thought it would be simple. I was told to go to the harbor in St. John's and look for the *Petrel V*. On the morning of August 4, I asked the taxi driver at the airport to take me to the harbor, and gave him the name of the ship. He called colleagues, asking where the ship was. He also called the coast guard – but no one had seen the *Petrel* in years. I thought I knew better and was dropped off at the harbor. The ship was not there. I looked for a phone booth and called RMST in New York and learned that it would be late.

This [delay] brought two unexpected nights in St. John's; I was unaware that the *Petrel V* was struggling with the engine. The bus ride to Grand Bank on August 7th lasted from 7:45 A.M. till 4:30 P.M. When we arrived, the *Petrel V* had just arrived. We had to load all the food by hand, including 3,000 liters of water for the *OV* and some Pepsi for George.

This was the first trip for the *Petrel V* in 5 years. The

crew of 13 was led by the 72-year-old retired Captain Michel Dusseau. The crew spoke only French (except the two lady officers Nathalie Godin and Caroline Bugeaud). The cook was Marc Pageau, and a sailor was Jason Dubé. Two guys were hired as ship's crew at Grand Bank, and they became local heroes. The local paper wrote about them after the expedition.

I was lucky to get a single cabin, but it was so stinky I doubted that I could ever sleep in there. This "cabin" was [really only] a bunk with a door, measuring about 1.9 x 1.3 meters. The *Petrel* left Grand Bank at 9:15 P.M. It turned out that the speed could not be increased due to engine problems, and we were going about walking speed. The engineer did his job and soon we could go faster.

Gunter Bäbler

Meanwhile, frustrations with the *OV's* positioning system and the failures of different parts of the ship began to wear on the planners.

The *Titanic* curse started that very first night. *Magellan* was slowly making its way to the ocean floor. All other expedition ships, including the *Petrel*, were at a safe radius away from us. And then our power went out. Completely. This was a catastrophe in the making for a couple of reasons.

First, the *OV* had to remain in one spot, almost on a dime, whenever *Magellan* was deployed. And *Magellan* could not be deployed if the seas and currents were too rough. *Magellan's* cable was extremely strong but it did have its limits. If seas were high and the *OV* was "bobbing about" on the surface with *Magellan* tethered a couple of miles below, *Magellan's* cable could easily snap. This is why *OV's* side thrusters were on autopilot, coordinated by GPS so that *OV* stayed stable on one discrete spot, not drifting hardly at all one way or another from that spot. With no engines and side thrusters, *OV* just started to randomly drift.

Second, without power, *Magellan's* decent was now in a free-fall. When emergency power came up on *OV's* bridge, we could hear the cursing and yelling coming from *Magellan's* control room. It was all a bit surreal. Emergency power bathed the bridge in that dull, red, *Hunt-For-Red-October*-disabled-submarine light you see in the movies. No one at the time knew what had happened or why it had happened. In fact, to this very day, I've never heard the explanation as to why all power suddenly went out that night. Not just for us . . . but for the *Petrel* as well.

We hailed the *Petrel* on our emergency channel and informed them that we had lost power, had no control of our position, and they should therefore pull as far away from us as quickly as possible to prevent a collision. To everyone's surprise, their captain responded that they too had just lost all power, were also adrift, and were unable to do anything until they got their engines back running. As I recall, it was a very long, dark, moonless night with everyone keeping their eyes peeled for an unfortunate drift or collision with another boat. As morning broke, we found ourselves with power renewed, not too terribly far from the *Titanic* site and luckily, also very much alone. Once we re-established communication with the *Petrel*, we found that they had somehow drifted a surprising distance out to sea. It took them almost the entire day to get back to the *Titanic* site.

Charlene Haislip

David Livingstone returned in *Nautile* from his dive to get a closer look at The Big Piece. Susan Wels was aboard *Nadir* and eloquently commented on his information briefing.

Now, we're all waiting to hear from marine architect David Livingstone, who went down in *Nautile* this morning for a close-up inspection of the piece. At 4:45 P.M., the sub returns, and David climbs slowly out of the hatch onto the *Nadir*. The area of the ocean bottom where The Big Piece sits, he reports, is full of steep cliffs and ravines. Parts of the seabed are as barren as a desert, while other areas look like hillsides blooming with sponges and other sea life. "The piece," he tells us, "is sitting vertically in the sediment. It appears to have suffered no damage at all in the past two years, and it looks superb. It's covered with the most marvelous shapes and colors, and all but one of the porthole windows are intact. The reflections in the glass," he adds," are quite amazing." Still, David admits it was a big disappointment to him that The Big Piece didn't begin to rise during his dive.

Susan Wels

We arrived at the wreck site on August 9th around noon. There was TV equipment on board needed by the *OV* crew and after two Zodiac rides, we continued to The Big Piece location. I did not want to miss the recovery of The Big Piece. After the first Zodiac left for the *Abeille*, I grabbed a lifejacket and pointed out that I wanted to take the next ride – and it worked out. Our Zodiac trip was a relaxing, slow ride. The recovery would take place next day. I almost overslept. Officer Nathalie woke me up just minutes before the last boat of the day left for the *Abeille*.

Gunter Bäbler

Profound, crushing disappointment hit everyone when The Big Piece didn't move from the ocean floor on August 9. David Livingstone, meanwhile, now thought

Our Story

> The Big Piece had come from *Titanic*'s port side, around cabin C-86. Subsequent examination of The Big Piece's hidden inner side later confirmed our 1996 belief that the piece had come from the vicinity of cabins C-79 and C-81, on the starboard side. We kept busy with a series of meetings about the upcoming television programs and procedures.
>
> Charles A. Haas

August 10 dawned clear, the blue skies and warm air lifting everyone's spirits on all four ships. "Today's the day!" replaced "Good morning!" as energy spread throughout the team. On the *Petrel*, the first officer Nathalie Godin shouted in her broken English to a group on deck, "The Big Piece, it begin to rise! [sic]" Similar cheers went up on *OV*. The Zodiac was busy taking the production crew and the historians once again to the *Abeille* from the other ships. Many of the team wanted to go, but the space was limited. The recovery crew on the *Abeille* prepared to receive this part of *Titanic* that had not seen the sun's rays since April 14, 1912. On this day, if all worked out as planned, that proud iron would see those rays once more.

> At 3:15 this afternoon, two orange lift bags finally broke the surface of the ocean. Immediately, the *Abeille Supporter* began maneuvering itself into position to pull The Big Piece in. At 5:40, the *Abeille*'s massive winch started winding in the cable, and one after another, the four remaining lift bags came up. At 6:20, a huge ring hung with chains rose out of the water, and then The Big Piece of the *Titanic* appeared – an enormous wall of steel, the sea smashing and spraying against its open portholes.
>
> Susan Wels

Bob Sitrick was chosen to dive with Christian Petron. They would examine The Big Piece as it lay extended under the lift bags and make a photographic record just prior to the lift.

> The plan worked perfectly, and I was fortunate enough to be asked to accompany our amazing underwater cinematographer, Christian Petron, down to approximately 100 feet and shoot stills of The Big Piece ascending to the surface. We were, therefore, the first people to see The Big Piece as it floated past us. As it bobbed on the surface I saw a dinner plate from the debris field was stuck to the side of the hull with rusticles. As The Big Piece was hoisted out of the water, it hit the side of the ship a few times and apparently jarred the plate loose. I was the only person to see it.
>
> Bob Sitrick

The *Abeille* team went to work. The main lifting line had to be secured to the ship's winch through open portholes on The Big Piece. George, PH, and Pierre all stood by, anxiously watching. Anyone could see the focus in their eyes as they observed the workers preparing the bed on which the Piece would be placed. George, in his shorts and sandals, would pace nervously back and forth, occasionally stopping to ask a question. After PH or Pierre answered, he would return to his erratic pacing. After all stations had double-checked their responsibilities, the approval was given, and the winch began to turn. Onlookers on higher decks stood holding cameras, riveted on the scene below. The energy of the crowd was evident in their expressions and in the articulations of hands, arms and bodies. Many of those watching didn't grasp the depth of emotion felt by George and PH. This was literally the moment these men had been waiting for over the past three years.

It was finally time.

The whine of the winch grew louder as it brought in foot after foot of lifting line. Crew members were overseeing the retrieval, but were cautious of the extreme tension in the line. When the Piece reached the surface, the process was paused to ensure all was ready on deck. Again, the winch began to turn, and this reddish-brown mass slowly emerged from the sea. The ship adjusted to the new weight from the A-frame, and the slight movement gently bumped the iron giant off the stern. Quickly the crew stabilized the Piece, and when the recovery officer was satisfied, the Piece was lifted completely out of the water. Those viewing from upper decks could not look away, except to occasionally glance at one another with grins. The iron giant glistened as the sun reflected off the wet beams. The ship still moved gently with the tide, creating a dynamic background — a slight motion to the sea behind The Big Piece.

George's pacing intensified; he wanted to jump in and help — but he forced himself to allow the team to do the job with which they had been entrusted. He stood alert, his expert eyes moving from the winch to the A-frame pulley to the connections holding The Big Piece. Recovery team members held the supporting cables taut as the A-Frame rotated inward, and the Piece was centered over the bed frame on the stern. Slowly, ever slowly, the winch was reversed and the Piece was lowered until lying flat. The Piece had been successfully recovered — safely, efficiently, and respectfully. There were hugs and handshakes throughout the crews and observers. George had asked that the group not cheer loudly out of respect for those who sailed on *Titanic*. The witnesses reserved their exuberance until later.

> As the process continued, I watched George overseeing recovery of The Big Piece. I could see that he was itching to help, to direct, to guide – but he yielded to the expertise of the pros aboard *Abeille*. I will never

forget his words, given quietly to each knot of spectators on the decks as the final lift began: "There will be no cheering, no demonstrations. This is a moment for remembrance, for history; The Big Piece is not a trophy."

Charles A. Haas

George was yelling at the *Abeille* crew to "Watch the ropes! Watch the glass! How do you say glass in French?" (George [was always reminded] that glass in French is *verre*.)

Cindy Tulloch

It was an incredible experience to watch The Big Piece ascend from the ocean floor and to actually touch it as it bobbed on the surface. It still amazes me that I was one of the first people to experience this, and that we were the first to see and touch that piece of the hull since the night the ship went down. Although it was nearly 20 years ago, I am still humbled and grateful to have had this experience.

Bob Sitrick

Watching the recovery of The Big Piece was definitely the high point. It was suspenseful, intense, and finally, very moving. I was grateful that I had a chance to come back and look at it later, after the flurry of activity in bringing it up and after the crowds had dispersed.

Steven Biel

There was an awed, stunned silence upon The Big Piece's first appearance above the waves. I found myself thinking it had been 86 years since sunlight had passed through those portholes. The awe yielded to momentary dismay when a sudden wind gust caused The Big Piece, now fully out of the water and suspended above *Abeille's* deck, to bump the crane frame, fortunately with no damage apparently resulting.

Charles A. Haas

Today, four portholes of the *Titanic* have risen high above the waves, for the first time since April 1912. They've done it. The Big Piece is up. George Tulloch and PH Nargeolet walked over to the hull section and together placed their hands on the cold steel. The *Nadir* whistled a salute. On the *Abeille*'s deck, where we stood watching, there was silence – only the groaning of the metal as the massive wall rose high above the fantail.

Susan Wels

On Monday, August 10 at approximately 5 p.m. (EST) the phone rang. George was on the line. The connection was faint, and I struggled to hear him.
 "Hello?"
 "Hello, Cin?"
 "Yeah, it's me George!"
 "Cin, we got The Big Piece!"
 "Really?"
 "Yeah, we got it. She's up on the *Abeille*."
They did it!

Cindy Tulloch

The Big Piece was secured to the stern, and the conservators moved in. Soon, a hose sent a gentle spray over the iron surface. If the Piece were allowed to dry out, the iron would become brittle and fragile. George walked around the Piece and would often squat down to touch it. He did not share his thoughts, but it was obvious his mind was reflecting back to 1912, to the time when that Piece was part of the largest moving object in the world; reflecting back to a time when the world was a different place, with different priorities and a different way of life. After his private moments with the Piece, he called for the others assembled together.

The historians began their close examination. The *Abeille*, at George's direction, moved to the other two ships parked over the bow some 15 miles away, and made a close pass, enabling everyone to see The Big Piece on the day of recovery. Along the port side of the *Petrel*, silent observers stood, many with cameras, some waving. Everyone knew the significance of the moment. As the *Abeille* passed, the *Petrel* saluted with her whistle. The *Abeille* then made a pass by *OV*, where three decks were lined with crew and production teams. *OV* saluted with its whistle as well, and those watching paid their respects.

A group sat on *Petrel's* deck to reflect. They stared at the *Abeille* sailing away, the angle of the evening sun creating golden sparkles on the wave peaks. "It's been a long time since that iron felt the sun," said one. The others in the small group nodded.

> The Big Piece is huge – 25 feet wide, 13 feet high – some 20 tons of steel plates milled in Scotland and riveted into the form of a ship's hull in Belfast. It is memory made physical, shaped from steel and glass. After 86 years and four months at the bottom of the ocean, its metal skin is mottled with black, rust, green and ivory-colored markings. Rusticles trail in clusters from its inside wall, which is striped with strong, vertical steel beams. It is still cold to the touch from the sea. It is a piece of the *Titanic*, and its steel remembers the last moments on the surface, the tearing of the hull and the long, dark fall down to the bottom.

Our Story

"Looking at this section of the hull, we can see that the ship's final moments were tortuous," says marine forensics expert William Garzke. "Its rivets are twisted like a wet towel. Its metal bearings failed. I think we can now say that the terrible noises that survivors heard were not the boilers exploding, but the final ripping apart of the *Titanic*'s steel.

Susan Wels

My memories of the recovery of The Big Piece are basically a lot of photos. When it finally broke the surface I could not believe how big it was, having been used to the other artifacts. I knew the dimensions, but The Big Piece appeared to be so much bigger than I imagined. When it was finally hanging in the A frame, the *Abeille* rolled and The Big Piece hit the frame. I felt the pain. It was as if someone had hit me with a hammer. This noise of the steel bending as it was placed down on the deck was also very powerful. So much steel was torn and ripped when *Titanic* went down in 1912, and that incredible noise described by the survivors was a tiny bit repeated as the Piece was placed down. I will not forget how everyone started to move in slow motion around The Big Piece. It was like a King Kong movie, where the "beast" is finally brought down and everyone walks slowly to make sure it is really dead and also to be amazed by it. At one point, the conservator (Olivier Berger) asked me to help, and we walked on The Big Piece on the inner hull, "inside *Titanic*" and we closed one of the large windows. What an incredible feeling to close a window of the *Titanic* after 86 years. It all worked so smoothly, with no resistance. I was amazed by The Big Piece and spent much time with it. It tells so many stories about the sinking.

Günter Bäbler

Late in the evening, Dave Livingstone and I went over some structural and arrangement plans to determine where The Big Piece may have come from. We decided it was indeed from the starboard side amidships. Later, we discussed this with Charles Haas and Jack Eaton.

Bill Garzke

While much of the attention was on the recovery of The Big Piece, on *OV, Magellan* began the reconnoiter dive. Eyes were focused on the monitors while a cautious ear was tuned into the conversation over the communication lines — the positioning system was holding strong.

Magellan successfully deployed on the seabed early in the evening. The ROV searched the three forward hatch openings for possible exploration attempts of the mini-ROV *Robin*. It appeared access for *Robin* would be problematic due to silt accumulation at the bottom of each hatch. Late that night, Tom Dettweiler, with my direction, began an exploration of the iceberg damage. *Magellan* gradually lowered starting at the main deck level at the forward end of the superstructure. Once reaching the line of sediments ploughed up by the bow encounter with the seabed, the ROV moved slightly along the exposed starboard plating. Tom and I finally were able to spot and film the iceberg damage that was a very small displacement on the inner plate inboard of no more than a half-inch. These sightings confirmed what Paul Mathias found in his sonar survey in 1996. The gap is very small and not easily detected. What made this different in 1998 was the lighting angle against the side of the hull. I was elated and said that this was a proud moment of his marine forensics career.

Bill Garzke

Also in the evening, a sudden storm hit before the teams were able to return to their base ships. *OV* was tossed about for several hours.

Things got more complicated when Zodiac runs between the four expedition ships were cancelled due to the bad weather, stranding 15 extra people on *OV*. People are sleeping everywhere – on the floor, on benches, wherever they can find space, while the vessel pitches and rolls throughout the storm.

Susan Wels

August 11, 1998

The next afternoon, once the storm settled, *Magellan* began its next launch. Production crews were excited about the vivid, clear images portrayed on the monitors. When the ROV arrived at the forward peak of the bow, Greg Andorfer talked directly with Troy Launay, who was piloting *Magellan*, to establish an "approach shot" to the bow. And they repeated it numerous times, from several angles, using different effects, in order to capture the perfect bow segment for the opening scenes of the documentary.

There is a difference between a scientific expedition and a television production. For this early phase of the expedition, science took the back seat. The numerous "bow passes" to get the perfect shot took almost two hours. If the producer saw an item that could be used in a documentary, the *Magellan* was asked to pass over and over it until the producer was happy with the capture. If there was something scientific to examine, it was noted for future dives. Those who were observing — the historians and general *Titanic* enthusiasts — wanted to see *Titanic*. There was so much to see. Instead, they had to watch these repeated captures of the same image over and over. It was frustrating, but there was significant time for conversation and collegial discussion. What originally was a time of boredom and repetition became a time of learning and sharing.

> It was back to the bridge to continue viewing images from *Magellan*. One point of interest was the large deformation in the hull plating on the port side of *Titanic*. Much later analysis of this ripple determined it was caused when the bow section struck the seabed and abruptly came to a halt. A storm came up later in the afternoon with 40-knot winds that limited discussions. *Magellan* was brought to the surface before the storm came to affect *Ocean Voyager*.
>
> Bill Garzke

Several members of the team had brought fishing equipment. One day, a school of Mahi-mahi came alongside the ship. They were large fish, some as much as four feet long, with bright colors and such a beautiful blue on the side. Those that were caught became dinner. The Malaysian cooks on *OV* served them with rice as this was the side dish for every meal.

> The NBC crew was so professional that they had time and mental space to kill fish and eat them.
>
> Saul Rouda

> I watched as several members of *OV*'s crew lined the rails, catching Mahi-mahi fish, about four feet long, with bright blue tops, silver underbellies and yellow tails. The fish attacked the lures within 20 seconds of their hitting the water. We had fresh fish for dinner that night.
>
> Charles A. Haas

NBC, on their *Dateline* show, would air the first segment, *Raising the Titanic*. This show was scheduled several days prior to the *Titanic Live* show and focused on the recovery of The Big Piece.

Meade Jorgensen came to me one afternoon and said an NBC affiliate had asked for an interview, and they would like for me to be at the starboard wing bridge to answer questions. I agreed, and arrived as planned. Meade set up the interview very well. He explained the affiliate would ask questions about The Big Piece, the expedition, and many topics, and that I should just be myself but try not to give away too many secrets about The Big Piece. We wanted the audience to tune in to the *Dateline* show the next day. "Tell how you feel, what you thought, what you saw," Meade instructed, "but don't tell TOO much."

The sea swells caused the ship to rock significantly. The background seen from the camera's perspective showed the horizon rising and lowering out of the frame behind me. With an earpiece, I stood on the wing bridge, holding on, and a soundman sat near me in a resin chair holding a boom microphone. Meade stood behind the camera, giving me the cue for a sound test. Shortly the link was made, and I was answering questions from a news team somewhere in the US.

One of the news team did not like the horizon behind me disappearing only to reappear again, but I explained this was the North Atlantic, and we had to work in difficult conditions on some days. Suddenly in mid-interview, the ship hit a rogue swell, causing us to rock side-to-side, and the resin deck chair, complete with sound man, began to slide toward the railing. The soundman was going overboard. I jumped out and grabbed him, pulled back, and stopped him from a fall into the ocean. "Are you ok?" I asked. "Yes. THANK YOU!" he answered with relief. Two others who were there watching helped him back to his original position. I walked back to the wing bridge, adjusted the earpiece and said, "I'm sorry – would you repeat the question? We just had a slight incident." The newsman was speechless. "Is everything all right there?" I answered, "This is a real expedition. Things happen all the time that we have to experience and deal with. And you get to witness this awesome moment!"

The news commentator tried his best to get me to give out secrets about the recovery of The Big Piece. My concluding remarks were: "We have a great team of people who have worked very hard on the *Dateline* show for tomorrow and the *Titanic Live* show that you can see a few nights after that. It would not be fair to them to tell any secrets – but I promise you this – those who watch will remember that show for a long, long time." I looked up, and Meade was doing the Rocky pose, both hands up in the air, pumping thumbs up.

A little later, the soundman told me that he was looking forward to getting back to dry land.

Bill Willard

That night, thunderstorms kept me overnight aboard *Ocean Voyager*, with Zodiac service to *Abeille* suspended. With no bunks available, I bedded down on an air mattress on the floor, under the tables in the dining room. The mattress and I slid back and forth as one, between the table legs, as the *OV* rocked. The next morning we learned that winds of more than 40 knots were encountered.

Charles A. Haas

On August 12, production was under way on *OV*, *Nadir* and the *Abeille*. Charlene Haislip had the immense task of coordinating "who goes where for what." At the same time, the NBC crews were assembling and preparing for the *Live* show to be aired only four days away.

David Livingstone and I boarded a Zodiac August 12 and were joined aboard the *Abeille* by Charles Haas and Jack Eaton to make a close-up inspection of The Big Piece. The structural piece revealed the terrible ripping apart that took place in the separation of the bow and stern sections. The section of the ship disintegrated during the hull failure and The Big Piece was the best evidence of the calamity that beset *Titanic*. A striking evidence of corrosion of 86 years on the seabed was revealed in the framing and brackets. There appeared to have been a 15-20% wastage in the steel. Another shocking failure was how C-Deck plating had ripped from its companion structure.

Bill Garzke

Back on the *Abeille,* I stood beside Bill Garzke as he examined The Big Piece closely. The steel's thickness measured three quarters of an inch where it had been in the ocean floor's sand; its original thickness was about one inch. The corrosive, acidic sand may have caused the thinning. More television interviews followed.

Charles Haas

The IFREMER team on *Nadir* observed strict safety guidelines. There was no debate or compromising those guidelines. To be in compliance for *Nautile* to dive, *Magellan* was not allowed at the same part of the wreck at the same time. If *Nautile* was on the bow, then *Magellan* must be on the stern section. The reasoning was simple. If for some reason *Magellan's* tether was cut or broken, it would fall to the ocean floor. Possibly over two miles long, this severed tether could hit and damage *Nautile* and trap that crew on the ocean floor with no chance of rescue. Again, plans had to be coordinated so that *Magellan*, with special high definition cameras from Woods Hole Oceanographic Institute, could capture extraordinary footage on the exteriors while *Nautile* with the mini-ROV *Robin* could begin to move into the interiors.

A daily routine was established, and the schedule was announced. Those being interviewed would be at a named location for filming at a certain time. *Magellan's* and *Nautile's* launch times were posted with target objectives for each. Producers coordinated which technical people would be at each site. Those who were "off duty" at any time could relax and enjoy sitting on the deck in the sun. Some read, and others engaged in conversations. When the cameras began to roll on the underwater vessels, quite a few made their way to the monitors.

> There were many down times when the ROV couldn't launch. I read Stephen King's 1,300-page book, *The Shining*. In fact, I read it twice during the entire six weeks.
>
> Bill Willard

> During down times, I read a lot. I brought along a novel by Henry James, *The Ambassadors*. It was one of those dense books that I probably wouldn't have read in another context – with a transatlantic theme even though it has nothing to do with ships or the sea, and published around a decade before the *Titanic* went down.
>
> Steven Biel

For the *Titanic* enthusiasts, however, the best place to be was in front of the monitors with peers and colleagues as image after image appeared from the ocean floor. The impromptu discussions and sharing of ideas bonded this group.

> Being in the control rooms, with the images of *Titanic* from below on the monitors, [was] fascinating. During my time out there, I met people with all kinds of

expertise: rusticles, submersibles, passengers' biographies, naval architecture, etc. I really enjoyed those conversations.

Steven Biel

August 14, 1998

As footage continued to come in, eager eyes absorbed views of each hallway, each room, each fixture and each nuance as *Robin*, and later, *Magellan* continued capturing *Titanic* as she was.

> Early on Friday, August 14, David Livingstone and I joined with Greg Andorfer to review the stern portion of the wreck. It is hard to believe that this once was a part of the ship. The hull plating was torn from the side transverse framing, and the well deck was bent over the poop deck, leaving a clearance of about 12 inches between them. The debris field around the stern is very chaotic with large sections of the ship ripped from the stern section by the phenomenon of implosion/explosion that occurs when air trapped inside the hull escapes when the pressure at a critical depth exceeds the strength of the plate. Close examination of some of the plates revealed that rivets were missing while in some areas the rivets remained but part of the plate was torn away. There were teacups, plates, serving trays, bedsprings, and utensils strewn about.
>
> Greg Andorfer had Tom Dettweiller maneuver *Magellan* around to the stern end that was intact to see if one could read *Titanic* etched onto the plating there. He wanted to prove that the wreck was *Titanic* and not *Olympic* as had been claimed by one book writer. Despite the efforts to find and photograph the name,

the amount of rusticles there aborted our attempts.

Later in the afternoon *Ocean Voyager* encountered rough Atlantic Ocean weather. The sea became very rough due to winds of 30-35 knots and five- to seven-foot waves. Heavy rain followed as a cold front moved through the area. Dave Livingstone and I met again in the evening to absolutely confirm the location of The Big Piece using the shell expansion plan. There is no question, based on their studies, but that it is from the starboard side of *Titanic*.

Bill Garzke

The camera moves slowly over floors covered with debris, around bent Corinthian columns, under crystal chandeliers suspended at weird angles, into a room whose walls are still covered with pale, carved paneling. For the first time, we are looking at the *Titanic*'s first-class dining room, through the eyes of *Robin*. A few days ago, the robot penetrated several areas of the *Titanic* that haven't been explored before – the first-class dining room, the doors of the three first-class elevators and the private deck outside Mrs. Charlotte Drake Cardeza's suite on B-Deck. Today, aboard the *Nadir*, Matt Tulloch and I are viewing two hours of videotapes that *Robin's* cameras recorded. Watching them, it's often hard to tell exactly where *Robin* is, because of the clouds of sediment and falling rusticles that frequently obscure the picture. Sometimes it looks like the red robotic eye is hovering in outer space, surrounded by what seem to be galaxies of swirling undersea debris. But then the image clears, and we're following *Robin* down hatches, farther down into the darkness of the Grand Stairway, through corridors, into the Marconi room and past windows of cabins that were once inhabited by the *Titanic*'s crew. It's remarkable footage, and much of it explores familiar territory for *Robin*. The robot has spent about 100 hours on the *Titanic* since it was first designed in 1985.

Susan Wels

As *Robin* entered Charlotte Cardeza's cabin, the viewing team reminded themselves of her steamer trunks that she carried as she traveled. The ROV crept in slowly, and someone pointed out the doorframe lying askew on the uneven floor, having broken off its hinges. With the doorframe in view, other characteristics of the suite began to be defined. As the particulates continued to move with the currents, extremely damaged wood wall paneling opened up into a heater insert, with a slightly identifiable mantle across the top. In the corners of the room, in vague shadows at the end of the ROV lights, a wicker chair was piled in the corner to the right of the image, and beyond the chair, what appeared to be the remains of a sofa. The ROV turned toward the passageway connecting the rooms of the suite. Quickly the ROV stopped forward motion. Impeding the way was a curtain of electrical wiring descending from the ceiling with no safe path through it. Viewing Charlotte Cardeza's steamer trunks would have to wait until another day.

- 7 -

The *Titanic* Live Program

IN THE FEW DAYS leading up to the *Live!* show, the entire expedition team settled in to routines. Everyone knew the production schedule, the dive schedule, who would be where, and who *had* to have an awesome Zodiac ride.

> Memories from the *Petrel*: The main bathroom was painted blood red and was located over the main engine room so it was akin to going pee in hell. It was bloody hot, morning, noon and night and really loud! Each morning we were greeted with blue and green large pancakes by the Quebecois chef, Cookie. Titanic pancakes were a staple aboard and not to be missed. I was very lucky to have a small room to myself with a bunk and sink. Although very close to the kitchen, it was a little private area to escape to! Laundry was always exciting as the small onboard washing machine located somewhere in the bowels of the ship would bounce around during the spin cycle while you tried to keep it from flying over. Of course there was no dryer so all the laundry had to be dried in your room or on

deck if you didn't mind everyone getting a good look at your unmentionables.

My favorite memory of being on the *Petrel* is the night and evenings. We would be away from the other ships, and they would turn off all of the lights on board so that we could look at all the amazing stars. It was fantastic! Most nights, Cookie would bring out his guitar and sing French songs as well.

Lori Johnston

I had a brief stay onboard the *Petrel* during one of the relocations for the *Live!* show. The cook on the *Petrel* was a short man with a big smile, jovial. Every morning he would say in his heavy French accent - "You want Titanic Pancake?" and he would make this huge pancake, larger than the plate, with a greenish tint and just laugh while cooking. But the man was a great cook! That *Titanic* pancake was a great memory.

Bill Willard

OV had its own share of living difficulties. The forward part of the ship had one large restroom. In it were four stalls with toilets, several sinks and four small shower cubicles each with a flimsy plastic shower curtain. With several females on board, there were awkward moments as one gender walked in while the restroom was in use by the other. The ladies began to go in pairs so that one could stand guard, or they would enlist a person who was idly sitting nearby.

In the galley were several games, of which Scrabble became the favorite. Jack Eaton and Charles Haas, along with an occasional challenger would be studiously into a game during idle times. Also in the galley was the lone television and VCR with a limited selection of B-grade movies.

Life on *OV* was its own small microcosm. *Nadir* had few common areas, and the *Abeille*, a work ship, had even fewer areas where people could get together and meet. Though *OV* was the central hub, each ship had its own areas where significant work was done.

The *Abeille* was the French vessel used for retrieval of The Big Piece. Following the retrieval, Dr. Cullimore dove the next day or so after, so during this time I was able to scoot over and do a full photo and retrieval of rusticles from The Big Piece. To preserve the integrity of the Piece, the conservators had seawater sprayed over it for most of the day, which did nothing to assist the photography or the retrieval, but oh well, life goes on. I wouldn't melt. I did however get soaking wet and thoroughly orange and red from rusticle retrieval.

I had done a full photo mosaic of both the top of The Big Piece (inside wall) and the bottom (outside wall) and even the piece of a broken dinner plate that was cemented to the bottom. This was the highlight of my rusticle career so far. The rusticles were absolutely amazing top to bottom, the variation of growth, the size, texture, colors and everything in between was an extraordinary experience.

After spending all day working, I was invited to dinner onboard. Meanwhile a storm was brewing . . . after dinner the sky was dark grey. It had started to rain and the water was quite rough. Okay, maybe my uneducated sea experience might exaggerate, but I was told that I would immediately be going back to the *Petrel*. I was unceremoniously put in a lifejacket and tossed into the Zodiac. The two Zodiac drivers were in full survival suits and thought that it was great fun to not only get me soaking wet (again) but to scare the pants off me as we flew over waves and crashed back to earth!! They succeeded nicely. I have never been so happy to see the *Petrel* come over the horizon and never climbed the rope ladder quite so quickly. Those onboard said that I had a lovely green tinge, so I promptly went to my room to die in peace. After a few hours, a nap and crackers, I was back in the saddle.

Lori Johnston

In order to get the NBC crews to *OV* and *Nadir* for the *Live!* show, numerous people had to change places. Many of us had to go to the *Petrel*, while the NBC team moved from the *Petrel* to *OV* and *Nadir*. The Zodiacs were working full time. Not only were people being transported, but they had to carry their luggage as well. We were moved soon after the *Petrel's* arrival so the NBC team could get started. On my day to transport, the waves were six to eight feet in height. We were bouncing along, the waves making for a fun trip. With me was an NBC soundman going to get his luggage and return to *OV*. As we pulled alongside the *Petrel*, they dropped the rope ladder. Several *Petrel* passengers were looking down on us. Julien Nargeolet instructed us to step off when the Zodiac was at its highest point, or it would be very dangerous. The ladder was moving up and down beside the Zodiac and the NBC soundman reached for the rung when the Zodiac was at its highest. His hands slipped. As the boat fell, he instinctively reached out again and grabbed a very low rung. In just a short moment, as the wave started back up, he would be underwater. I lunged forward and wrapped my arms around him with an "I've got you!" He began to go underwater until he released the rung. As the Zodiac rode the wave, I was able to pull him in. We were both soaking wet, but we knew he was going to have to try again. A couple of waves later, he successfully locked on to the higher rung and was cheered as he climbed aboard.

Bill Willard

The *Titanic Live* Program

August 15, 1998 – The day before the *Live!* show

Work was non-stop and at warp speed on *OV*. NBC teams met, discussed, edited and refined plans for the next day. Planning, spoken in a foreign language of technical terms and trade talk, was about minutes and seconds and setting up segments. It was intense, but most impressive to observe. On *Nadir*, it was time to attempt the first-ever live link from the ocean floor.

> If a delicate and risky experiment goes as planned this afternoon, we may have the first live transmissions in history from a manned submersible 2.5 miles down at the bottom of the North Atlantic Ocean. It's a technological feat that has never been attempted, and it involves considerable danger to the submersible's crew.
> Here's how it will work (if it works):
>
> - On Wednesday night, the *Abeille Supporter* dropped a basket carrying 6,500 feet of fiber-optic cable into the Atlantic, positioning it some 5,000 feet from the *Titanic*'s bow.
> - Late last night, the *Abeille* made another drop – this time, a metal cage connected to a live fiber-optic cable uplink to the ship. At 5:45 this morning, the cage was successfully positioned some 2,600 feet equidistant from the *Titanic*'s bow and stern.
> - This afternoon at 1:45 p.m., the *Nadir* crew launched the manned submersible *Nautile*, which is now on its way to the bottom. When it reaches the seabed, *Nautile* will retrieve the fiber-optic basket and carry it in its robotic arms to the live uplink cable in the cage.
> - *Nautile* will then try to connect itself to the fiber-optic cable and attempt to connect the cable to the live uplink.

- If all the connections are successful, TV monitors on the *Abeille* will immediately begin receiving live color video images from *Nautile's* four cameras, positioned outside and inside the sub. Those images can then be transmitted by microwave to the *Ocean Voyager* and from there sent instantaneously via satellite to the world.

"If this technology works," says George Tulloch, "it will enable people everywhere to experience, real-time, the wreck of the *Titanic* and the reality of the deep ocean. And they'll be able to participate, real-time, with the people who are actually risking their lives to explore it."

Now, a word about the dangers:

First, the long fiber-optic cable could get jammed inside *Nautile's* rear propeller or wrapped around the sub's robotic arms. If that happens, *Nautile* will have to cut itself free from the cable and make an emergency return to the surface.

Second, if the *Abeille Supporter* is unable to hold its position or has to abandon the site because of rough weather, the live uplink cable could spiral down on top of *Nautile*, trapping those inside.

Third, if the ship moves, the uplink cable could develop a kink, which will make it useless. The $200,000 cable will have to be cut loose and lost.

"We understand the risks," George says. "That's why our team has worked for more than two years studying and dealing with the dangers of placing a manned submersible near an uplink cable. To reduce the risk, we've engineered safety clamps and lines to catch the cable in case it breaks at the surface. At least, that will give the *Nautile's* crew a chance to escape."

As I'm writing this in the galley of the *Abeille Supporter*, we've heard that *Nautile* has reached the seabed and has successfully located the fiber-optic basket.

5:30 P.M.: I'm crammed inside a TV production shed on the *Abeille Supporter* with George, PH Nargeolet, photographer Olivier Pascaud and members of the NBC Titanic crew, staring at a bank of video monitors. On one screen, we watch a black and white video image of the sandy sea bottom, shot from a camera on the uplink cage. Suddenly, the screen shows *Nautile* flying slowly in deep ocean toward the cage like an alien spaceship, dangling the basket filled with fiber-optic line.

Watching the sub approach in black and white is like watching the 1969 moon landing. I don't think it would surprise any of us to see Neil Armstrong climbing out of the hatch and taking one small step onto the ocean floor. But this isn't the moon's Sea of Tranquility. It's the bottom of the earth's Atlantic Ocean, and all of a sudden, it's coming to us live. *Nautile* places the fiber-optic basket on the seabed, stirring up a huge cloud of ocean-bottom dust. Slowly, the haze of sediment recedes, and we watch one of *Nautile's* robotic arms reach out to the cage and grab the connector that will hook the fiber-optic cable to the *Abeille's* live television uplink.

Minutes go by as *Nautile's* R2D2 arms delicately plug the fiber-optic cable into the connector. Then, in an instant, the TV monitors fill with crisp, color video images from the *Nautile's* cameras.

Unbelievable. The robotic, bottom-of-the-world live video connection worked. Even Pierre Valdy, its inventor and chief engineer, doubted that the transmission could succeed. But it has, and the results are almost better than anyone had hoped.

Susan Wels

The day's activities also included a dry run of the documentary with everyone involved. NBC's Bob McKeown introduced everyone and reviewed the schedule. Since the show would be live, a dress rehearsal was planned for 10:00 P.M. in the dining area (7:00 P.M. New York City time), at which time everyone was informed of their specific roles in the show.

> The chief "heads and beds" person, Charlene Haislip, assigned me to *OV's* cabin 30, with large windows on two sides and a private bath – nice! I felt sad that Jack was still in the tiny *Abeille* cabin. Starting at 11 P.M., we did a run-through of the live telecast set for the next night, including questions that I would be asked and the program's sequence. David, Bill, and I were all wedged together opposite correspondent Bob McKeown, but the bridge also had a soundman, two camera operators, one executive producer, one production assistant, one assistant producer, a deck officer, etc. Despite the crowd, it was difficult to maintain footing due to *OV's* rolling. Late at night, I returned to "my" room to find NBC's Phil and Curtis already in it, behind a locked door. Instead of a large, luxurious suite all to myself (dream on!), I slept on the hard floor with only a pillow, using my sweatshirt as a blanket.
>
> Charles A. Haas

August 16, 1998 – The day of *Titanic Live*

The stage was set, the actors rehearsed and ready. Four ships on the North Atlantic full of technicians and experts in many fields were in place. In New York, and the MSNBC studio, a backup team was in place for interviews along with a scripted show in case everything on the ocean failed. In the studio was Tim Foecke, who had performed numerous metallurgical tests on the metals recovered

The Titanic Live Program

from *Titanic*; historian Claes-Göran Wetterholm, forensics analyst Dick Silloway, and ROV co-creator Susan Willard. Each had prepared answers for questions should complications form at sea. Filmed segments were available for the studio also.

As the time approached to go live, the stresses began to rise and peak. Intermittent weather squalls created an issue; could *Magellan* and *Nautile* launch safely? If the launches were delayed, how would this impact the live broadcast? The *OV* office was filled with the leaders of the different groups involved. Oceaneering's team leader George Brotchi was asked, "Can *Magellan* launch?" George's response was, "Can the ship stay on station?" PH felt that *Magellan* could launch in the current conditions, but had concerns about a launch if the conditions worsened.

The producers sat with scripts in hand, awaiting direction for the show. George Tulloch, in his always-positive attitude, was attempting to convince everyone that all would be fine, and to prepare to launch. The decision on whether to go live or transfer operations to the studio was in the hands of Maureen Lemire. She made the call.

> I had never done live television before. Of course NBC had the responsibility of the "live" elements and had such an impressive team. We, the Discovery team, besides working on the science show that would air some months later, handled much of the role in material that gave color and backstory. These segments could be used as fill or save the live show if it had any technical difficulties, which a storm on the horizon was threatening. But my live responsibilities were largely going to be watching the feed go out, the vision mixer switching from amazing location to location – topside anxiety and maneuvers to the live hookup and images from below!
>
> I was on the *Ocean Voyager* where the *Magellan* was deployed. It provided the critical powering line for the live signal from the ocean floor. I checked in with the

bridge as the *Magellan* was going down. Our broadcast was to start within the hour.

   The captain was unhappy. He had been preparing for the live event with his favorite Bordeaux. He nonchalantly declared the weather conditions were no longer suitable for us to have a tethered ROV off our stern. It had to come up. It was a bit choppy and we had aborted deployments before – but we were going live, soon! This was our moment! The captain didn't care. He simply stated he could not hold the ship steady. Given his condition, I wondered if it was just him. As the ship gently swayed, I looked over at the second in command, Captain Tony Foster, and asked for a quick, panicked, word. Could he hold the ship steady in these conditions? No pressure but the whole live event is riding on this. Tony said yes, without hesitation. I then executed my most critical expedition call as the senior ranking Discovery employee on the ship and relieved the captain of his duties. Tony slid into the captain's chair. All my power went to my knees as I quaked an exit from the bridge. Captain Tony's kids got an extra bag of Discovery swag and letter about how their dad saved the day. And he did.

Maureen Lemire

A rush of adrenaline and busy preparations filled the day. I reported to *OV's* bridge at 10:15 P.M. and got wired with a microphone, then took several photos. The program began at 11 P.M. our time (8 P.M. New York time), and despite two thunderstorms passing overhead, it went remarkably well, except that I had to use both hands to cling to the edge of the chart table to prevent myself from tumbling into Bob McKeown before a worldwide audience of millions, as the *Ocean Voyager* rocked during the storms. It was a wonder my white knuckles didn't show.

Charles A. Haas

It was the big day for the television special on NBC. There were practice runs scheduled for this morning, but were canceled in favor of a further development of the script. Paul Mathias came over from *Nadir* and was chosen to go down in the *Nautile* and maneuver *Robin* around in the Reciprocating Engine Room. There were some intermittent squalls that raised some concerns about underwater activity later. The script was altered again.

After dinner, Dave Livingstone and I went to the quarters of the regular master of *OV* after being told by PH Nargeolet to remain there while Tony Foster, the alternate master, would be at the helm to keep *OV* on station. Dave and I brought him a bottle of wine and chatted with him with sincere sympathy concerning his demotion. He was quite philosophical of his role tonight. We left him in good spirits.

Bill Garzke

The Master of *OV* looked like Keith Richards – and acted a lot like him, too.

Troy Launay

After the walk-through, I wandered from station to station. I had no specific duties during the show, so I could observe from different locations behind the cameras. There were a select few of us who watched in this way. I chose to watch Bob McKeown. Bob was to go on camera in the *OV* office, and a nice backdrop and foreground was prepared for him. Bob's notes were on the table before him, and he went over the notes meticulously. During the previous few hours, Bob and others were snacking on a bowl of Jolly Ranchers in the office area. Bob's flavor was cherry. When the call came from the location producer with a 10-minute countdown, Bob, as did all the others, checked his hair, clothes, teeth, and then he noticed, with horror, that his tongue had a huge red streak down the middle. The Jolly Ranchers had left a souvenir! Bob quickly tried to rinse the red away, used a toothbrush to scrub the red away, and nothing worked! Exasperated, and running short on time, he took out a small compact – there were no makeup specialists onboard. "He's going to do it," I said aloud. And he did. He took the powder puff, stuck out his tongue, and with one brave sweep tried to coat his tongue. His face proved it was not a fun experience, and when he looked in the mirror, discovered it was not a successful attempt. I stepped in to offer assistance. "Try the yellow ones," pointing to the Jolly Ranchers. "They may tone down the red." Bob immediately nodded, opened two and popped those into his mouth. It didn't work, but no one noticed the bright red streak!

Bill Willard

In the NBC trailer, Phil Alongi carried the scepter of power. His technicians at the controls had worked together before and everyone knew the procedure. In layman's terms, the director — Phil — would instruct the team as to which camera angle to show on air, and those under Phil's leadership were informing the announcers to stand by, that they were up next, to introduce the next segment, and much more. It was an orchestration with numerous dynamic parts, with live announcers Bob McKeown and Sara James, with Paul Matthias in *Nautile*, with recorded segments queued and ready to play. Back at MSNBC, the backup team watched in the green room, ready if needed, and John Siegenthaler bracketed the commercials and returned the network to the NBC trailer where the team was all set for the next segments.

As the time for the show neared, the squalls had subsided, and the sea became wonderfully calm. This pleased George Brotchi, and the earlier tension transformed into a "now it's time to do what we do best!" attitude — not haughty or arrogant, but powerfully professional. The *Live* show proceeded from segment to segment, interview to interview, footage to footage. And then, the world saw live images from *Titanic* for the first time.

> *Titanic Live* commenced at 2300 (ship time) and lasted until 0100. The show went on without any difficulties and many contributed to its success. Paul Mathias contributed his part some 12,000 feet below while a thunderstorm passed *OV* by around mid-broadcast. The public was also shown the recovery of The Big Piece, as it was brought safely aboard ship and soon would begin its journey to Boston harbor. Finally after 86 years, a piece of *Titanic* would finally make her entrance to North America.
>
> Bill Garzke

> The most emotional parts of the trip for me were first seeing the live feed from the manned submersible when it plugged itself in and then hearing our host say "You are looking at something that nobody has ever seen before . . . ." Over the course of more than a year we had been planning for that moment, and a tremendous amount of brain power and hard work had gone into making it a reality. I will never forget those two moments in time, and the goosebumps they raised up and down my body.
>
> Bob Sitrick

At the site, those aboard the *Petrel* had no feed, no way to see the masterpiece come together on *OV*, *Nadir* or the *Abeille*.

> During the live show I was stuck on the *Petrel* due to the breakdown of the Zodiacs. There was not a single ride

in 60 hours from the *Petrel* and I was stuck there! I was upset, as the world could watch but I was within sight of the ships and could not see it. I hung out with Lori Johnston and Steven Biel. We had a great time and enjoyed BBQ, etc. Interestingly, on the morning after the live show, suddenly three Zodiacs approached us (for days none were available)! The reason – our two lady officers attracted the sailors of the other ships – it's all a question of priority to make a Zodiac work!

Gunter Bäbler

After the conclusion of *Titanic Live*, there was a big celebration by all hands involved in the show's production. NBC personnel broke out a large store of beer and liquor from their van that dwarfed the items brought aboard by the Discovery and RMS Titanic personnel. A big party went on to the wee hours of the morning.

Bill Garzke

Right after the program, I fielded radioed questions forwarded from New York for about 45 minutes, with my answers being sent out via the Internet, then a brand-new experience for me. To celebrate the program's successes, we then had a "wrap party" on *Ocean Voyager's* helicopter deck. I went to bed (the floor!) at 4 A.M.

Charles A. Haas

In New York, the studio had a small party as well. The group at the studio would depart the next morning for St. Johns, by way of Halifax.

Dick's role started at the MSNBC studio during the *Live* show. He went prepared, and he was disappointed that he didn't even get to say anything or answer any questions. He was just there. That backup team wasn't [utilized].

Beverly Silloway

The results were a resounding success – for Discovery the second-highest-rated program in their history and for the Titanic exhibit – The Big Piece. For me, it was the first of what would be several broadcast firsts in my career, but still is and will always be the trip and experience of a lifetime.

Bob Sitrick

~ 8 ~

Intermission
The First Phase Ends

The *Live* show was finished. NBC reported ratings as soon as they had numbers. The television people on the expedition cheered as they were announced. The crowded galley was filled with laughter, music, handshakes, and hugs. As the Zodiac runs brought more and more people to *OV*, greetings of congratulations rang out as the team arrived from the other ships. When PH and George entered the room, both men responded to the cheers with big smiles. It was good to see the happiness on everyone's faces now that the stress was gone.

Paul Matthias sat near the corner of a table, sharing his perspective from *Nautile*. He had been a late arrival to the party, as had the Oceaneering crew; they had to bring *Nautile* and *Magellan* back onboard and secure them before joining the gala.

Slowly the crowd diminished. In the morning all four ships would depart the site. The NBC crew would head to Ireland to cover President Clinton's visit. The *Abeille's* destination with The Big Piece was Boston. *Nadir* and *OV* would meet in St. John's, resupply, and return to the site for the second half of the expedition.

INTERMISSION: THE FIRST PHASE ENDS

The next morning, team members were up early, preparing to return home or on to their next assignments. Some would not be returning to the site so they said their goodbyes.

On this overcast morning, there was a lot more traffic than usual in our floating city on the sea. Zodiacs were cutting across the grey water from one expedition ship to another, transferring passengers and gear to and from the *Abeille Supporter*, the *Ocean Voyager*, the *Petrel V*, and the *Nadir*.

Susan Wels

For those returning home, Maureen Lemire gave each a jacket commemorating their participation in the 1998 Expedition. It was a memorable voyage. David Wood and Dick Silloway would come aboard later representing the Marine Forensics Panel for the second portion of the Expedition.

Bill Garzke

I returned to Boston on the *Abeille*. We brought The Big Piece with us.

Steven Biel

I was aboard the *Petrel* for the trip to St. John's. The weather was warm and scant clouds dotted the sky. About noon on the first day, someone noticed a pod of dolphins off the starboard bow, very close to the ship's peak. They were racing the ship, just as seen in Cameron's movie. They were majestic as they became airborne, then dove gracefully back into the sea, all the while kicking at full speed. We watched for a long while until they turned back toward the open sea.

Bill Willard

Our Story

> Jack arrived on board the *Ocean Voyager* on a Zodiac from the *Abeille*. We were told we would transfer to *Petrel V* for the trip to St. John's. But in response to a serious concern expressed by Tom Bethge, the expedition's safety officer, Jack refused to budge from *OV*, which was the right thing to do. By radio, *OV*'s captain already had filed an entry passenger manifest with Canadian authorities, so he definitely was not amused about an unexpected passenger. But I treasured the memory of Jack's triumphant appearance on the deck of *OV*, smiling and waving as I climbed the 40-ft rope ladder of the 50-year-old rust bucket, and then we were off for Newfoundland. I was assigned to an inside cabin with no porthole, just a sputtering air conditioner. I shared the cabin with conservator Olivier Berger.
>
> Charles A. Haas

Over the next two days, *OV* began a return voyage to St. John's in Newfoundland. At one point 36 miles from St. John's, those aboard experienced what it was like to be a fisherman off the Grand Banks. A heavy fog blanketed the ship and reduced visibility to less than 50 feet. The ship made only a cautious eight knots, but it was a reminder of those days when speeding passenger ships collided with unsuspecting fishing boats in dense fog conditions. As *OV* made her way into St. John's, we entered the harbor in bright sunshine.

Bill Garzke

We've just finished the first leg of the '98 Titanic expedition . . . . This evening . . . the ocean is nearly as quiet as a lake. A few hours ago, we saw the dark tails of pilot whales breaking the water's surface off the bow. Above us now is a clear sky filled with falling stars, as we head at a steady 10 knots north toward the Grand Banks.

Susan Wels

Intermission: The First Phase Ends

The journey on the *Petrel* became legendary to the new roster of passengers aboard for the first time. As the ship rocked and rolled toward St. John's, more of the team found special moments to remember.

> Olivier Berger saved my life early this morning. He came back to the cabin at about 1:15 A.M. after having finished watching a videotape of the movie *Independence Day* in *Petrel's* "lounge." He was just dozing off when he heard a loud bang and saw a two-foot flame erupt from the electrical outlet into which the air conditioner was plugged near my bed. He immediately pulled the plug, using his socks to protect his hands. I slept through the entire episode. Had he not acted, fire would have blocked our path to the door – the only exit – and the room, under many layers of paint, would have gone up quickly. Ever since, each time I see *Independence Day* on television, I give thanks for Olivier Berger's quick thinking and saving my life. Had Jack accompanied me as my *Petrel* roommate, the outcome might have been very different.
>
> The ship's electrician came to repair the electrical box, which had burned out, leaving scorch marks that reached the ceiling. The room became very hot and airless due to the air conditioner's absence. I moved my flashlight near my bed in case there were other emergencies.
>
> Signs on board *Petrel* tickled all of us with their tangled language and forthrightness: "No more lundry [*sic*] please, we are low on water," and "This toilet jams very easy. Use another if a lot." Another bore the legally required words "To lifeboats," but inexplicably pointed downward. I spent much of the day reading, but was happily distracted by seeing more dolphins and female whales with their calves.
>
> Charles A. Haas

Our Story

Aboard the *Petrel*, passengers gathered on the open deck, enjoying the sun. *Nadir* and *OV* sailed ahead, while the *Petrel* did its best to keep moving forward. As it approached St. John's, ahead was an immense fog bank concealing land. The dense fog was short-lived, and as the ship emerged from the bank, a majestic panorama appeared. A narrow channel led into St. Johns' harbor. To the left, a World War II gun emplacement could be seen still guarding the entrance to the harbor, a monument to the city's history. To the right of the channel was Signal Hill, with a communications station atop it. Passengers and crew lined the decks, absorbing each moment. The *Nadir* and *OV* were already docked, and *Petrel* pulled up behind them. A large crowd cheered the ship as it arrived, townsfolk waving and celebrating the successes of the *Live* show.

Having arrived during the previous night, the crews of *OV* and *Nadir* had the fortune of viewing the harbor by night.

> Suddenly, at 10:00 P.M., we see the blinking beam of Cape Spear lighthouse through the fog, and then the lights of a small city gleaming in a cleft of hills. Everyone is strangely quiet on the bridge as we approach the coast. Now the red and green lights of a pilot boat appear, and the pilot, John Wakeham, comes aboard to guide the *Nadir* into St. John's narrow harbor. We slow to four knots, then three, as we pass Esso oil tanks and the steep rise of Signal Hill, where Guglielmo Marconi received his first transatlantic wireless transmission in 1901. The *Nadir*'s captain, Michel Houmard, shouts orders as he maneuvers the ship deftly into port. At 10:45 P.M., the *Nadir* pulls up dockside, and we all crowd the rail to get a good, long-awaited look at solid ground – as well as at the fifty people and two kilted bagpipers who have unexpectedly arrived to greet us. "We saw you on the news," John Wakeham tells us, "and we all thought the *Titanic* was coming in. Welcome to St. John's," he says, "the City of Legends."
>
> Susan Wels

Intermission: The First Phase Ends

As each ship arrived they were met at the dock by a small crowd of *Titanic* fans. The dock crowds waved and took photographs. As everyone departed the ships, hands were shaken, questions were asked and words of congratulations were shared.

> When we arrived in St. John's, we were happy to see Cindy Tulloch and Florence Nargeolet, who were joining the expedition's Phase Two. While we were in port, vague rumblings began about Tropical Storm Bonnie, forming in the Caribbean. But the storm was more than 2,000 miles away.
>
> Charles A. Haas

The ships' crews ventured into the city. Those on the *Petrel* transferred their gear to *OV*, and checked in with Charlene Haislip as to where their new cabins would be. The expedition team separated into various groups, each going shopping, dining, touring, or just to have a friendly libation.

> Early in the evening, a problem developed. The NBC trailer and equipment were scheduled to be removed from *OV*, and the ship was to be moved to a construction dock. The ship's master was away in town, and the officer in charge gave permission for the ship to be moved. I returned early, and being the first person back, could not find the ship. After a few minutes, Larry Daley, shipping agent in St. John's, arrived and showed me the ship some distance around the harbor. In a short time, I was driving his van, picking up expedition members and transporting them to the ship.
>
> Late in the evening, someone said, "There is a problem." The ship's master had arrived, somewhat inebriated, and was upset that his ship had been moved without his permission. He demanded I take him to the coast guard office, where he was going to have

> whoever was responsible arrested. As we neared the ship, he calmed down, and two of the crew promised to get him to his quarters safely. At 2 A.M. local time, I parked the van, and Larry Daley took me to the ship. By the time I arrived, the construction crew had removed the cargo trailer.
>
> Bill Willard

During day two of the St. John's stop, a plane arrived with the three in-studio guests from the *Live* show. Dick Silloway, Claes-Göran Wetterholm, and Susan Willard arrived for the second leg of the expedition, along with several members of the Nauticos team. Day three would be the final day on shore. The ship would be restocked and the NBC equipment removed.

> Rather abruptly this evening, at 9:30 P.M., our shore leave was declared over. Tonight, we'll start heading back on the *Nadir* to the *Titanic* site, and the *Ocean Voyager* will follow us tomorrow once some repairs are made. No other vessels will be joining the Titanic expedition for the second half. The *Petrel 5* has ended her voyage in St. John's, and the *Abeille Supporter* will soon be pulling into port in Boston.
>
> Susan Wels

Expedition Photos

Our Story

Onlookers lined the decks of two cruise ships in 1996 to watch the first attempt to raise The Big Piece. Photo courtesy of Günter Bäbler, and RMS Titanic, Inc.

Light towers used to illuminate the wreck in 1996, designed and built by Christian Petron. Photo courtesy of Günter Bäbler, and RMS Titanic, Inc.

EXPEDITION PHOTOS

Ocean Voyager during the 1996 Expedition. Notice the empty helicopter deck. Photo courtesy of Günter Bäbler, and RMS Titanic, Inc.

Ocean Voyager during the 1998 Expedition. Notice the extra trailers on the helicopter deck and the fantail. Photo courtesy of Günter Bäbler, and RMS Titanic, Inc.

The MV *Abeille Supporter*. Photo courtesy of Charles A. Haas, and RMS Titanic, Inc.

The *MV Petrel V*. Photo courtesy of Günter Bäbler, and RMS Titanic, Inc.

EXPEDITION PHOTOS

The IFREMER vessel *Nadir* hosting the submersible *Nautile*. Photo courtesy of Christian Petron, and RMS Titanic, Inc.

A *Nautile* launch is underway. Photo courtesy of Christian Petron, and RMS Titanic, Inc.

Our Story

The Oceaneering ROV *Magellan 725*. The two aluminum cylinders in the front house a high definition camera system from Woods Hole Oceanographic Institute. This was the first time high definition images were taken at the site. Photo courtesy of Günter Bäbler, and RMS Titanic, Inc.

The fantail of *OV* with *Magellan*'s crane and equipment trailers in view. The small door on the starboard side of the ship is the entry/exit point for Zodiac transfers. Photo courtesy of Günter Bäbler, and RMS Titanic, Inc.

EXPEDITION PHOTOS

George Tulloch entering a Zodiac from *OV*. The sea was calm on this day. Photo courtesy of Claes-Gören Wetterholm, and RMS Titanic, Inc.

Zodiac transfers were a fun part of the expedition. With large ocean swells, the ride was more adventurous. Photo courtesy of Charles A. Haas, and RMS Titanic, Inc.

With the *Petrel* alongside *OV*, a Zodiac transfer is taking place. Notice the rope ladder being lowered for the passengers. Photo courtesy of Charles A. Haas, and RMS Titanic, Inc.

During moderate swells, PH Nargeolet waits for the Zodiac to be at its highest point before pulling himself up the knotted rope. Max Salmhofer looks forward, watching for the next wave. Photo courtesy of Charles A. Haas, and RMS Titanic, Inc.

EXPEDITION PHOTOS

A Zodiac leaves *OV* towards *Nadir* on calm seas. Photo courtesy of Claes-Gören Wetterholm, and RMS Titanic, Inc

Angus Best, expedition geologist, examining his core sampling equipment. Photo courtesy of Charles A. Haas, and RMS Titanic, Inc

"Garzke and Hutch" – Bill Garzke, forensics analyst, and David Livingstone, architect from Harland and Wolff, examine The Big Piece. Photo courtesy of Charles A. Haas, and RMS Titanic, Inc.

Author Steven Biel, on deck during down time. Photo courtesy of Günter Bäbler, and RMS Titanic, Inc.

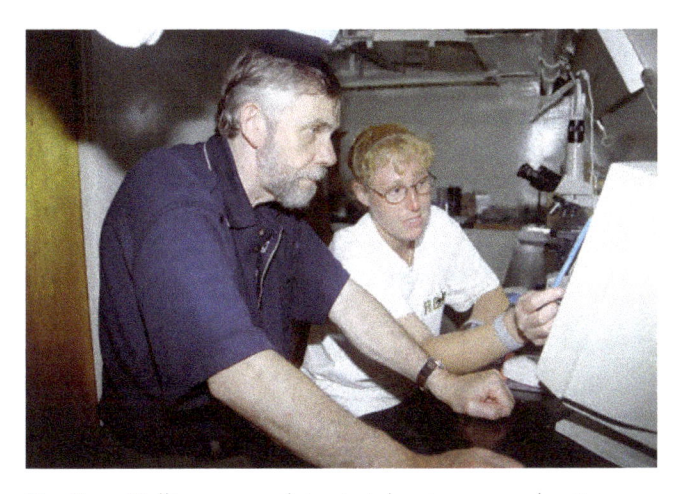

EXPEDITION PHOTOS

Dr. Roy Cullimore and Lori Johnston, conducting an analysis of rusticles and iron-eating bacteria on the wreck. Photo courtesy of Günter Bäbler, and RMS Titanic, Inc.

Expedition physician, Dr. Robert Budman, complete with pirate eye patch. Aaargh! Photo courtesy of Claes-Gören Wetterholm, and RMS Titanic, Inc.

Our Story

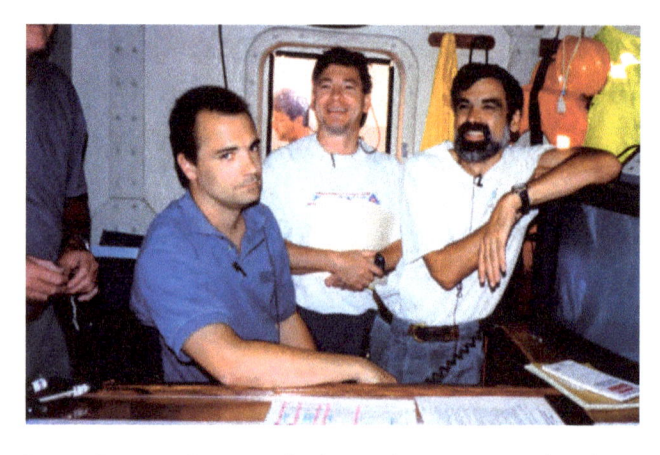

Bruce Brown, Greg Andorfer and Tom Dettweiler share a funny moment while working. Photo courtesy of *Nauticos*, and RMS Titanic, Inc.

John P. "Jack" Eaton, historian, author and Scrabble champion. Photo courtesy of Günter Bäbler, and RMS Titanic, Inc.

EXPEDITION PHOTOS

Charles Haas, Bill Willard and Susan Willard pose for a post-Bonnie photo with *T-Rex*. Photo courtesy of Claes-Gören Wetterholm, and RMS Titanic, Inc.

Mark Knobil, camera and Saul Rouda, sound, for Stardust Visuals. Photo courtesy of Claes-Gören Wetterholm, and RMS Titanic, Inc.

The Scrabble Championship is under way! George Tulloch takes on Jack Eaton in this round. Photo courtesy of Charles A. Haas, and RMS Titanic, Inc.

Bob McKeown, Charles Haas and Bruce Brown relax at the conclusion of the *Live* show. Photo courtesy of Charles A. Haas, and RMS Titanic, Inc

EXPEDITION PHOTOS

Meade Jorgensen, executive producer of NBC's *Dateline* show. Meade is holding a congratulatory cigar, celebrating the success of the *Titanic Live* show. He has a seasickness wristband and watch on each wrist. One watch shows ship's time, the other New York time. Photo courtesy of Meade Jorgensen, and RMS Titanic, Inc.

Susan Willard, Bill Willard and Claes-Gören Wetterholm pose for a post-hurricane photo on the fantail of OV. Photo courtesy of Claes-Gören Wetterholm, and RMS Titanic, Inc.

Charles Haas and Jack Eaton celebrate successful entry into a Zodiac. Photo courtesy of Claes-Gören Wetterholm, and RMS Titanic, Inc.

Dick Silloway, Tim Weihs, and David Wood study footage. Photo courtesy of Charles A. Haas, and RMS Titanic, Inc.

EXPEDITION PHOTOS

The women of OV, second phase of the expedition L-R: Susan Wels, Charlene Haislip, Florence Nargeolet, Susan Willard, Stephanie Ratcliffe, Cindy Tulloch. Photo courtesy of Charles A. Haas, and RMS Titanic, Inc.

Members of the team study monitors showing footage from the ocean floor. Photo courtesy of Stephanie Ratcliffe, and RMS Titanic, Inc

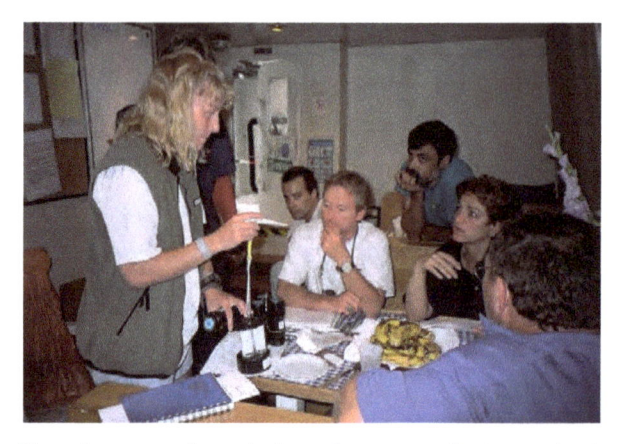

Planning meeting during the second phase. Lori Johnston explains the BART test to Bruce Brown, Gary Hines, Tom Dettweiler, Charlene Haislip and George Brotchi. Photo courtesy of Stephanie Ratcliffe, and RMS Titanic, Inc.

Dr. Roy Cullimore goes over specifics. Lori Johnston and Gary Hines are left of him, with David Wood and Angus Best to the right. Charlene Haislip, David Elisco and Tim Weihs are on this side of the table. Photo courtesy of Charles A. Haas, and RMS Titanic, Inc

EXPEDITION PHOTOS

Bob Sitrick and Christian Petron were chosen to dive and photographically document The Big Piece during recovery. Photo courtesy of Christian Petron, and RMS Titanic, Inc.

The first lift bags raising The Big Piece arrive at the surface! Photo courtesy of Günter Bäbler, and RMS Titanic, Inc.

The Big Piece from *Abeille* as it breaks the surface. Photo courtesy of Günter Bäbler, and RMS Titanic, Inc.

The upper half is out of the water, as seen from a Zodiac with the divers. Photo courtesy of Bob Sitrick, and RMS Titanic, Inc.

Expedition Photos

The Big Piece from the Zodiac. Photo courtesy of Christian Petron, and RMS Titanic, Inc.

On the *Abeille*, diver Bob Sitrick inspects The Big Piece upon its arrival at the surface. Photo courtesy of Günter Bäbler, and RMS Titanic, Inc.

An up-close photo of The Big Piece showing rivet holes and many different patterns of rusticle growth. Photo courtesy of Günter Bäbler, and RMS Titanic, Inc.

Rivet pattern on one side of the Big Piece. Photo courtesy of Günter Bäbler, and RMS Titanic, Inc.

EXPEDITION PHOTOS

The site of a porthole, long broken away with what appears to be fibers on the metal's surface. Photo courtesy of Günter Bäbler, and RMS Titanic, Inc.

The three men who raised The Big Piece: George Tulloch, PH Nargeolet and Pierre Valdy. Photo courtesy of Günter Bäbler, and RMS Titanic, Inc.

The historians present on the *Abeille*: Günter Bäbler, Charles Haas, Steven Biel, Jack Eaton Photo courtesy of Günter Bäbler, and RMS Titanic, Inc.

Sara James of NBC, recording a segment of *Raising the Titanic* for Dateline NBC. Photo courtesy of Günter Bäbler, and RMS Titanic, Inc.

EXPEDITION PHOTOS

George Tulloch in an interview with Sara James, NBC. Photo courtesy of Günter Bäbler, and RMS Titanic, Inc.

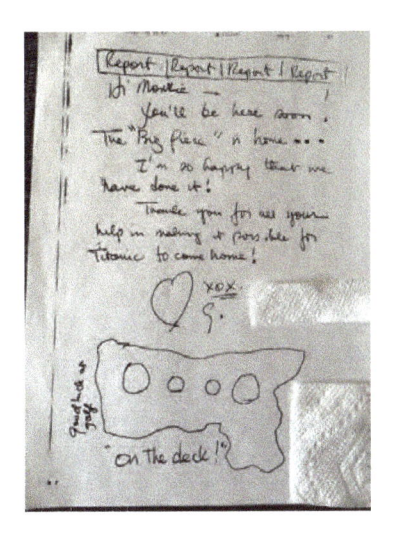

Fax sent from George Tulloch to his wife Cindy letting her know The Big Piece was recovered and safely on deck. Photo courtesy of Cindy Briggs Tulloch, and RMS Titanic, Inc.

The Big Piece has been recovered, and is on it's way to Boston. Photo courtesy of Günter Bäbler, and RMS Titanic, Inc.

The first hours of Hurricane Bonnie. Photo courtesy of Nauticos, and RMS Titanic, Inc.

EXPEDITION PHOTOS

There is one majestic part of hitting a large wave with the bow — it makes an incredible splash! Photo courtesy of Nauticos, and RMS Titanic, Inc.

Hurricane waves wash over *OV*'s fantail. Photo courtesy of Cindy Briggs Tulloch, and RMS Titanic, Inc.

As *OV* goes up a steep wave, water floods the fantail because the rear of the ship is under water. The flooding worsens as the angle of the ship increases during the storm. Photo courtesy of Cindy Briggs Tulloch, and RMS Titanic, Inc

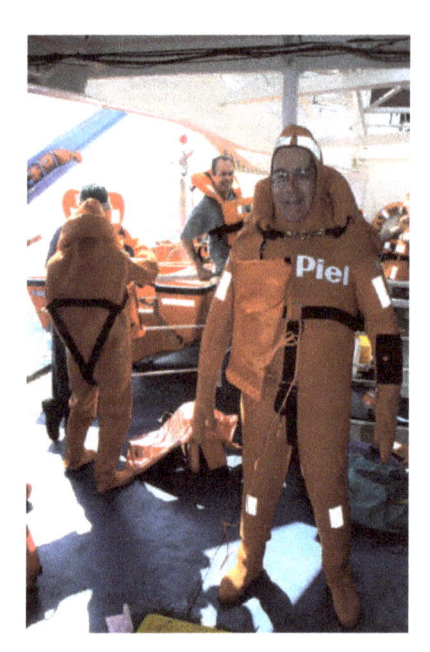

Charles Haas models a neoprene "Gumby" survival suit. Photo courtesy of Claes-Gören Wetterholm, and RMS Titanic, Inc.

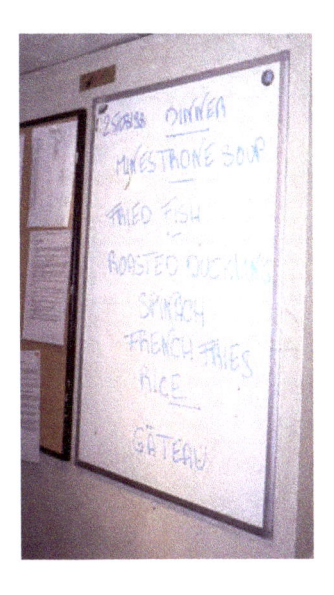

Expedition Photos

For each meal, the menu was posted on this board. Lunch and dinner ALWAYS included rice! Photo courtesy of Claes-Gören Wetterholm, and RMS Titanic, Inc.

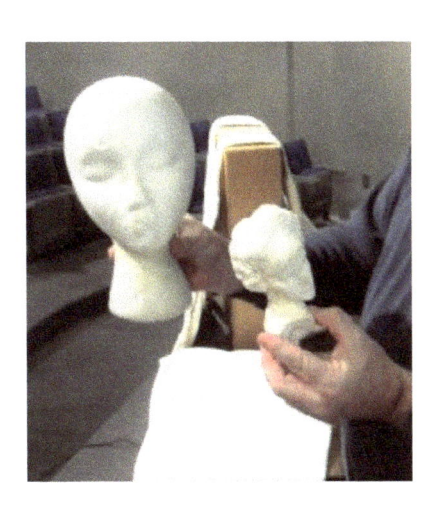

Styrofoam wig forms, such as on the left, were sent down to the immense pressures at *Titanic*'s depth. It returned compressed, as shown on the right. Photo Courtesy Anderson Indepent Mail.

Our Story

Styrofoam drinking cups compress to thimble-sized souvenirs. Photo courtesy of Maureen Lemire.

Tiny petrel birds that fit in your hand dotted the decks every morning. Photo courtesy of Charles A. Haas, and RMS Titanic, Inc.

148

EXPEDITION PHOTOS

The NBC control room for the *Live* show. Photo courtesy of Günter Bäbler, and RMS Titanic, Inc.

T-Rex launches to explore *Titanic*'s bow. Near the horizon, *Nadir* is visible, having launched *Nautile* to explore the stern. The distance between the two ships is the approximate distance between *Titanic*'s bow and stern on the ocean floor. Photo courtesy of Nauticos, and RMS Titanic, Inc

T-Rex enters the water, followed by *Magellan*, for a 2.5-mile descent to the ocean floor. Photo courtesy of Nauticos, and RMS Titanic, Inc.

Magellan's control booth. Photo courtesy of Günter Bäbler, and RMS Titanic, Inc.

Expedition Photos

Two artifacts recovered from the debris field: a container for sugar and a bowl as recovered. Cleaning would come before stabilizing each piece for transport to the conservation lab. Photo courtesy of Claes-Gören Wetterholm, amd RMS Titanic, Inc.

A thick glass window in its frame recovered from the debris field. Photo courtesy of Claes-Gören Wetterholm, and RMS Titanic, Inc.

View from St. John's dock of a fog bank protecting the harbor entrance. The *OV Zodiac* takes a cameraman to capture the image for the documentaries. To the upper left is Signal Hill. Photo courtesy of Claes-Gören Wetterholm, and RMS Titanic, Inc.

St. John's in the morning, as seen from Signal Hill. Photo courtesy of Claes-Gören Wetterholm, and RMS Titanic, Inc.

EXPEDITION PHOTOS

Crewman Jimmie James hoists a White Star burgee for the last hours on board, as *OV* sails into Boston. Photo courtesy of Claes-Gören Wetterholm, and RMS Titanic, Inc.

George Tulloch's business card could have included additional titles such as Visionary, Encourager, and Motivator. Photo courtesy of Cindy Briggs Tulloch, and RMS Titanic, Inc.

Expedition team members during the second half of the expedition, September 1998. Photo courtesy of Charles A. Haas, and RMS Titanic, Inc.

The Second Phase

Expedition Members on the Second Phase

Ocean Voyager
Cabin

1	Lori Johnston	
1		
3	Stephanie Ratcliffe	
3		
5	John Coombs	
5	Mark Vicuna	
6	Robert Budman	
6		
8	Matt Taylor	
10	Jack Eaton	
10	Charlie Haas	
11	Ron Schmidt	
11	Troy Launay	
12	Stan Jandura	
12	Frank McKinney	
14	Bill Willard	
14	Susan Willard	
15	Tim Weihs	
15	Claes-Göran Wetterholm	
16	Gary Hines	
17	Mike Picariello	
17	Brian Twomey	
18	Roy Cullimore	
18	Angus Best	
19	Saul Rouda	
19	Mark Knobil	
20	Tom Dettweiler	
20	Bruce Brown	
21	Julien Nargeolet	
21	Max Salmhofer	
25	Dick Silloway	
25	David Wood	
28	George Brotchi	
28	Chris Russell	
29	Florence Nargeolet	
29	Cindy Tulloch	
30	Charlene Haislip	
30	Brandon Plonka	
31	Steve Ulrich	
31	David Elisco	

Nadir

Olivier Berger
Muriel Boucherat
Charlie Burnham
Deuce Dubois
Paul Matthias
PH Nargeolet
Olivier Pascaud
George Tulloch
Matt Tulloch
Susan Wels

~ 9 ~

The Second Phase Begins

August 19, 1998

*N*adir left St. John's followed by *OV* the next day. Onboard were new members of the team, historian Claes-Göran Wetterholm, scientists Tim Weihs and David Wood, forensic analyst Dick Silloway, ROV co-creator Susan Willard, and Cindy Tulloch and Florence Nargeolet, wives of the expedition leaders.

> I travelled via Halifax [where] I visited the different *Titanic* cemeteries, this time accompanied by Susan Willard. There was a *Titanic* gift shop and a Dawson cafe, everywhere reminders of the *Titanic*. Incredible. I assume this is all gone today.
>
> Claes-Göran Wetterholm

> We had a house on the lake at the time of the expedition. One evening, we were having dinner on the deck, and Dick said, "I got this phone call today – it seems they want me to go on this Titanic expedition." He didn't think it was legit – he thought someone was

playing a practical joke on him. We all agreed. Dick spent the next 10 days researching, to determine if it was a legitimate request. Finally, he found out it really was the Discovery Channel who called him. He wanted to know who had recommended him! He was reluctant at first. Our family laughed about this for years!

Beverly Silloway

I remember the beautiful day and surroundings when we departed from St. John's. There was lots of excitement as we met each other; the weather was great – sunny but not too hot. I also remember boarding the ship and seeing all the crates that awaited me. They contained metallographic equipment, on loan from the Buehler Company, for preparing and polishing *Titanic* plate steel and rivets. I was excited to get underway and set up the equipment. Once we left harbor, though, and the coast began to fade, the excitement was tempered just a bit, when we received lessons in how to enter our survival suits. The reality that our voyage might have some risks began to settle upon me, particularly with the cold water. Little did I know what lay ahead. Long before the voyage, I remember sitting in my office at Hopkins and talking with Tim Foecke at NIST [National Institute of Standards and Technology] about the voyage. He had been asked to go, but given he does not like being at sea, he asked me to go in his place. He said that the small waves on his lakes in Minnesota were as much as he wanted to handle. Anything bigger made him nervous. Thus, I received the offer of a lifetime.

Tim Weihs

We marveled at the spectacular scenery as we left St. John's aboard the *Ocean Voyager*, passing through a fog-shrouded cleft in the cliffs sheltering the harbor, and heading out into the Atlantic. We loped along at about six to eight knots because one propeller was not working. As shipboard life settled down, we met and came to admire the talents and expertise of the team's latest members: Bill and Susan Willard, Angus Best, Tim Weihs, David Wood and Dick Silloway.

Charles A. Haas

We got our cabins on the *Ocean Voyager*, quickly renamed The Roller Voyager, all according to her behavior in water. My sea legs are non-existent and this ship became a constant reminder of it.

Claes-Göran Wetterholm

After the first leg of the expedition, the *Ocean Voyager* became my new home. As one of the few women aboard, there was confusion about where I was to bunk. It didn't bother me where or with whom as I realize life on board is cramped and limited. After much shuffling and one specific refusal for me to bunk with them, one of the crew graciously gave up their room for me. It was a lovely little room down with the crew, which were a great bunch of guys and often kept me laughing day and night. The one drawback about the room is that it was right over one of the bow thrusters, so while on station and the bow thruster was functioning, you could not hear a thing the noise was so loud. So at night, I would have earplugs in to sleep, however that meant that I couldn't hear my alarm. This was solved by one of the crew coming to wake me up each day, poor them. The other drawback to being in this small room was the washroom and shower facilities . . . there weren't any. I was able to use Dr. Cullimore's bathroom during the day, and did find a shower of sorts that the divers used to rinse equipment. After some equipment shifting and piling, I was able to climb over and use the shower!! Life is never dull!

Lori Johnston

Our Story

Once at sea I was able to setup the saw, the polishing wheel, and the microscope to capture early images of the rivets that we were recovering. The setup gave me something to do as we traveled to reach the wreck site and then waited for the search vessels to launch and recover materials. I should note that the Buehler Company had loaned us state-of-the art metallographic equipment, and it was fun to use. I owe them a good bit of thanks as well. I was able to capture a few images of rivets fairly quickly, until news of the approaching hurricane halted progress. I needed to repack all of the equipment to make sure it would be safe in the storm. I did repack after the storm but most of the analysis was done after we returned to land, back at Johns Hopkins University and NIST. Buehler was kind enough to let us continue to use the equipment back at Hopkins.

Tim Weihs

The Great Scrabble Tournament began, with several games each day, with varying participants, but Jack Eaton nearly always was the victor. We had another lifeboat drill, and this time we had to get into our immersion suits without any assistance. What a struggle!

Charles A. Haas

Jack Eaton and Scrabble – Jack would put down a word that most of us had never heard of before. It would always be a valuable play, of course. When called on it, "Jack, there is no such word!" He would straightly respond, "Of COURSE there is!" then proceed to give the part of speech, and one, sometimes two definitions, and even use it in a sentence if we were not convinced! There was no dictionary onboard to try and prove him wrong – several of us looked for one!

Bill Willard

The Second Phase Begins

Now that the stresses of the *Live* show were over, it was back to the business of science while making documentaries. *Nadir* arrived on site, and as soon as practical, deployed *Nautile*.

> All the while at sea, we monitored weather reports from both NOAA and those received by *Nadir* from France's IFREMER. We were looking for hurricanes that always made their way this time of year from the Tropics, moving northward offshore along the US east coast, all the way up to the northern latitudes right around the *Titanic* site. If we were lucky, we'd get our full 30 days at the site and not have to shut down early, trying to outrun a hurricane. Meanwhile, *Ocean Voyager* tended to [develop] daily mechanical problems. Its non-functioning desalinator was a terrible inconvenience. But consistent engine problems and the inability to have its GPS control the thrusters [created difficulty throughout the entire expedition]. [Again, we had to rely on] manual control by our oft-inebriated captain.
>
> Charlene Haislip

> Once the Oceaneering vehicle was down by the wreck site, we spent many hours watching live images of the ship. It was rather breathtaking to see this massive and historical wreck up close. The images of the ship were beautiful and helped us understand the damage. As a metallurgist or materials scientist, it was very exciting to see the examples of deformation and fracture in the steel, as well as the corrosion that was ongoing. Most striking, though, were the images of the smaller, personal items – shoes and luggage. These brought home the tragedy more directly and kept us remembering all the folks who lost their lives.
>
> Tim Weihs

> We spent hours viewing tapes and footage of *Titanic*'s current state, identifying targets for further study. Reports about now-Hurricane Bonnie were not reassuring; it was rapidly intensifying, with 110-knot winds, but still far off. We were told that our safety was of paramount importance.
>
> Charles A. Haas

Some of the best moments on the trip did not involve television schedules or ROV launches. On the ocean, as far out as *Titanic* is from the shore, there are massive amounts of birds. They are small birds, called Petrels — the exact name of one of the ships on site. These birds are gentle and docile. At night they are attracted to the lights on the ship and often will hit the ship while flying toward the lights. The birds land on deck, but do not have enough take-off room to get airborne on the short decks. Early in the morning, it brought great calmness to many to walk out on deck, at any door, and look for the birds. Several people would pick a bird up, spend quality time with it then gently toss it overboard. The birds would naturally sense their home environment, and would soon be soaring in the sky. Each day, the deck had 20 to 30 birds, minimum — plenty for everyone to enjoy for a few minutes of the morning.

August 21, 1998

> Tonight, at 8:00 P.M., the winds are whipping at 22 knots, and the waves are swinging up like a solid grey wall behind the *Nadir's* fantail. The ship is rolling and diving on the foaming chop like a carnival thrill ride. And *Nautile* will be breaking the surface any moment. "Tonight it's gonna be a wild one," George Tulloch predicts.
>
> George looks noticeably concerned. PH Nargeolet

is in *Nautile* today, having spent the dive investigating the *Titanic*'s Marconi Room and a door lying in the debris field. George stands on the bridge as the sky darkens with storm clouds, watching for the sub to emerge in the grey curls. A minute later, we see a small shape bouncing in the chop. *Nautile* has surfaced off the bow, and the Zodiac recovery crew will have to tow it to the *Nadir's* stern in the raging weather. A few more punishing moments, and *Nautile* is finally on deck and locked into her cradle.

Susan Wels

Have you tried to do physical work on a platform that rises and drops 10 feet every six seconds? Most of the team had not experienced ships continually responding to nature's waves. Some mornings were stressful as the major decision makers met to decide if the weather allowed safe deployment of ROVs and the submersible. Some days the sea was calm, on others, everyone held on as storms passed. To add to the excitement, an approaching hurricane motivated many individuals to radically alter perspectives. Each day's weather brought a new challenge.

One pretty rough day, my shooting crew had to transfer, and we put all our many bags of equipment into garbage bags for protection. Each crewmember, with a more experienced seaman on each side, was pushed into the Zodiac. Lastly a bucket brigade threw all our stuff into the rubber boat. The reverse process of boarding the next ship, was equally terrifying. After we were safely on deck with shaky knees, we realized we had more garbage bags than we started with. A few of them turned out to be . . . garbage.

Mark Knobil

Our Story

August 23, 1998

This morning at 10:30 A.M., Polaris Imaging's Paul Matthias and *Nautile* pilots Max DuBois and Yann Houard climbed into the sub and began their slow descent to the *Titanic*. Today, for the first time, they were going to begin digitally photographing the wreck site in order, ultimately, to create a high-resolution photomosaic of the *Titanic*'s bow and stern and the surrounding area.

"Our goal for the rest of the expedition," Charlie Burnham explained, "is to take approximately 100,000 pictures, then mosaic them with a computer and end up with as complete a photographic map as possible of the *Titanic*'s wreck. For the first time, we'll produce an image that's large enough to enable people to see the entire site, yet high-resolution enough to clearly show individual details."

It's a risky proposition, though. No one has ever tried to use two of these high-resolution color cameras at once, on a sub, to photograph a wreck in the deep ocean. And, as Max DuBois points out, it can be especially tricky using equipment that needs a lot of electrical power – like strobe lights – in the water.

So today, as often happens out here in the middle of the North Atlantic, things didn't go as planned. One strobe light failed on the way down to the bottom, and the other short-circuited after only 100 images of the debris field had been shot. At 5:00 P.M., Paul, Max, and Yann climbed out of the sub onto the *Nadir* looking unusually stressed and grim. Things looked slightly better, however, a few hours later. The 100 photos that they took today, even with one strobe light, looked pretty good when Deuce [Dubois] pulled them up on his Silicon Graphics workstation.

Now, at 11:00 P.M., Paul, Charlie, and Deuce are back

at work on the *Nautile*, repairing the sub's imaging equipment. Max, for one – a veteran sub pilot – takes a philosophical view of today's setback. "It's not surprising to have problems when you do this kind of work," he reflects. "After all, the sea is very hostile to men and to equipment."

Tomorrow, they'll give it a second try.

Susan Wels

August 24, 1998

Despite the bad-tempered rampages of Hurricane Bonnie in the south, the weather today in the North Atlantic is bizarrely tropical. Summer's almost over, but today the sun is Caribbean-hot, and the ocean is shimmering like a lake of silver. It's been a grand day for the Polaris team, too. At 5:45 this evening, the *Nautile* crew was back on board after a successful imaging dive. Today, everything worked right – the high-resolution cameras, the strobe lights and the computers. In almost four hours on the bottom, Paul Matthias and pilots Patrick Cheilan and Xavier Placaud managed to shoot 2,000 color digital images of the *Titanic*'s stern as well as the debris field, flying the sub low and slow 15 feet above the wreck.

This evening, in the computer shed above the fantail, Paul shows me a few of the color images that they took. The detail and clarity is amazing. In one shot of mangled wreckage from the stern, I can see a little strip of brass, three inches long, with three tiny holes for screws. It's so clear that it looks like I could reach into the computer screen and touch it – but the real object is two and half miles below me, in total darkness on the bottom of the ocean.

He shows me another picture, this time of a wheel

from the *Titanic*'s engine room. The wheel is covered with slick rust-colored corrosion, and clinging to one of the spokes is a ghost-white galathea crab. Because the image is digital, Paul can zoom in on details, and in seconds, the crab – which lives in perpetual night on the seabed – is appearing life-size, brightly lit, on his computer screen. "That's part of the beauty of these digital images," he says. "The resolution is extraordinarily high, and we can zoom in very close on the details and end up with a lot of useful information." Once the photomosaic of the *Titanic*'s stern is complete, Paul tells me, he could theoretically enlarge the entire image to life size. We could end up with a virtual stern section of the hull, 350 feet long and 90 feet wide, with every single detail in its place.

Susan Wels

Before the hurricane arrived, several of us went out on one of the Zodiacs to take a swim. The experience is locked into my memory for two reasons. One is that the water was relatively warm. The Gulf Stream was clearly over the wreck site, bringing us warm water that made swimming very comfortable. The warmth of the water and the ease with which we swam made me think of the many passengers who fell into frigid waters and did not last very long. Given the Captain's desire to reach the US in seven days, they would have avoided the Gulf Stream and hence would have been in much colder waters. The survival rate would have been much higher if they had been traveling within the Gulf Stream, even for that time of year.

The other reason the memory of the swim is very clear is due to a sighting. While swimming among the three-foot swells, we saw a dorsal fin about 100 feet away. Once sighted someone yelled shark and we all swam as fast as possible back to the Zodiac. Needless to say our hearts were pounding. Only after getting back into the Zodiac did we realize there was more than one dorsal fin and what we feared to be a shark was actually a small pod of dolphins coming to see who was swimming in their waters. We were all quite relieved and had a good laugh over our mistaken fear.

Tim Weihs

August 25, 1998

The night had been horrible. My cabin was in the bow, and the sea was rough. As soon as the ship shifted, the bow thrusters worked like mad, the whole cabin was shaking and there was no chance to sleep. Not a good way to begin an expedition.

Claes-Göran Wetterholm

August 26, 1998

On the 26th I noted there was an attempt to recover one of the landing doors but the current was too strong and there was too much wreckage blocking a safe recovery. However, I noted the sea was incredibly beautiful with large waves and the sun making a glittering road across the sea. I noted in my diary there is worry about the hurricane, and I just prayed to God that it wouldn't come close. But how little did I know!

Claes-Göran Wetterholm

Even the sea is strangely stagnant – flat out to the horizon and heaving rhythmically, like the respiration of a sleeping animal. Something fearsome may be coming. If weather reports are right, it may be a category-3 hurricane – powerful enough to cause extreme damage, both on land and to ships at sea. Since August 22, the *Nadir* has received warnings about Hurricane Bonnie, which started brewing in the Caribbean. Over the past four days, she's gained in strength and violence – her winds whipping up from 65 to 100 knots and her gusts climbing from 80 to 120 knots.

Today, Hurricane Bonnie started slamming the North Carolina coast. If she catches a low-pressure trough, she could quickly swing northeast and tear across the North Atlantic, bringing 50- to 60-knot winds and 25-foot seas between August 28 and September 1. We are right in Bonnie's path if she turns and heads for the Atlantic.

Susan Wels

George Tulloch showed me a 1996 image of an entryway to a third class stairwell on the stern – "Billy, what do you think about this?" asked George. I looked at a dive plan I had created from *Titanic* maps created by Charles Haas and Jack Eaton. "That stairwell leads to quite a few third class cabins. Imagine what we could learn from those rooms." We talked at length about many possibilities, not only of the cabins, but of the general meeting room close to the stairwell. Sadly, the *Magellan* flyover of the stern revealed a fallen piece of framing that partially blocked the entryway. "Let's put this on the list of objectives for 2000," George said.

Bill Willard

The dives continued. *Magellan* deployed daily, as did *Nautile*, weather permitting.

The strongest memory, though, was waking up in the middle of one night to catch our first glimpses of the third section, a large piece of the bottom that had broken off from the much larger bow and stern sections.

Being upside down, it offered us the first ever views of the bottom of the ship and showed how unscathed it was. The massive bending of the bottom plates, where it had broken off from the bow and stern sections, also demonstrated to me that the steel plates were ductile and did not behave in a brittle fashion as some newspapers described at the time. There were clear 90-degree bends of the bottom plates. And of all the steel plates, these bottom plates, that formed the bottom section of the double-hulled ship, would have been the coldest ones on the ship. Thus, if any steel would have behaved in a brittle fashion these would have and they clearly did not.

Tim Weihs

Also onboard was Dr. Roy Cullimore whose Biological Activity Reaction Test (BART) tests were deployed in several specific areas of the wreck. The BART tests were platforms containing pieces of metal with varying degrees of stress used to observe and measure the bacterial growth and rate of decay.

> . . . [Dr. Cullimore] would have liked to spend more time on the *Nautile*, so he could get a better look at the *Titanic*'s stern. "It's a mangled mess of steel," he explains, "that shows the tremendous forces at work when the *Titanic* sank." What especially intrigues him is the rate at which the ship seems to be deteriorating. Dr. Cullimore estimates that the wreck is losing a tenth of a ton of iron every day to iron-eating rusticles, microbial communities that have colonized the ship. He predicts that the *Titanic* has only 20 years left before it suffers what he calls a "biological implosion" and collapses as its infrastructure fails.
>
> His main mission is to learn more about the rusticles. Today, to investigate their feeding preferences, he

left samples of 15 different types of steel under the *Titanic*'s engines in the stern, where they'll remain for an entire year. "Rusticles are unusual biological formations," he says, "because they are consortiums of different microbes, growing and cooperating in ways that we simply don't see on the surface of the earth."

Susan Wels

Science definitely was at the forefront, with lots of experimentation and observation, including Paul Matthias's photo-mosaicing of the entire wreck site, the placement of Roy Cullimore's biological activity detectors, and Angus Best deploying his core sampler.

Charles A. Haas

There finally came a point where one of *OV's* two engines just plain died. We informed Discovery that it would cost $15,000 and take a couple of days to get the needed replacement part from St John's to the *Titanic* site. Instead of someone immediately "calling in the order," Discovery, RMS Titanic, and *OV's* owner all started arguing as to who was responsible for this expense. Days ticked by with no one willing to step up to the plate. We all thought this was somewhat comical until Hurricane Bonnie was birthed and started to make her way up toward the expedition site. We were shocked that an approaching hurricane was not enough to make someone pay for the part and then argue about reimbursement later. But it was not. The finger pointing continued to the very end when we were forced to abandon the site, charting a course back to Boston, straight into the hurricane, with only one engine functioning.

Charlene Haislip

August 27, 1998

> On the 27th, I noted I was deeply impressed by the incredible pictures from the *Magellan*. It was a completely new experience to be on the deck of a ship and almost four kilometers below, a camera gave us wonderful images that were crystal clear.
>
> Claes-Göran Wetterholm

> The *Magellan* ROV was an incredible tool for exploring the *Titanic* and is still in use today. There are very few ROV systems that explore at 7,000 meters.
>
> Troy Launay

The launch of the mini-ROV *T-Rex* was scheduled for this day. The helicopter deck hosted a small group, including Bill Willard, Susan Willard and others witnessing the first dive. The Oceaneering team powered their crane, and lifted the *Magellan* out of its cradle. *T-Rex* was cradled in a basket beneath it along with 300 feet of neutrally buoyant tether connecting it to *Magellan*. This umbilical would send images from the *T-Rex* to *Magellan*, which would transmit the digital signals through its three-mile tether to the surface. At the same time, in the control booth of the *Magellan*, operational control signals would go from the booth back to *Magellan* and on to *T-Rex*. The images from the *T-Rex* camera were displayed on *Magellan's* auxiliary screen in the booth. The Willards designed *T-Rex* for the purpose of entering small openings, revealing images from areas yet unseen inside the ship.

As the "go" was given to launch, the crane rotated, and slowly the *T-Rex* basket sank beneath the surface, then the *Magellan* unit itself entered the water. With several system checks, the order was given to deploy, and the cable began to unspool, *Magellan* disappearing from view. After the long two hours or more to drop, *Magellan's* pilot, Troy Launay, was at a loss. *Titanic* should have been there. It wasn't. A call to the bridge revealed *OV* had moved due to a "staying on station" problem. Launay was not happy. The bridge relayed coordinates, and *Magellan* began to move, but still no *Titanic*. Launay made the call to reverse direction. Eventually, slow movement brought *Magellan* closer, and *Titanic* came into view.

The initial test of *T-Rex*, however, was not as effective as everyone had hoped. The thrusters and power system were designed for a small, lightweight unit. It was discovered later that changes had been made in the system by someone other than the Willards, who designed and created the ROV. These changes tripled the mass of the unit, but thruster modifications were not updated to accommodate the mass change.

The mini-ROV did record images of debris in the proximity of the lowering cage, but was unable to venture deep into the interiors of the bow. PH arrived after his dive in *Nautile*, and George with him, and they watched and witnessed the conversations over the *T-Rex's* complications.

> George motioned at me for us to step outside. We went to the galley, where we had privacy at the time. "What has happened with *T-Rex*?" he asked. We talked for a long while about the unauthorized changes. George became very upset to know that neither of us, the Willards, were allowed to examine our own system to see what changes had been made in the circuitry. "Can you fix it?" he asked. I responded, "I don't know what is wrong right now. I'd like to have time to see what they've done." George took a couple of deep breaths. "Billy, there is a storm coming. I don't know if we will have a chance this trip to go down once more. Will you have it fixed for 2000? I want *T-Rex* in those third class cabins. Can you have it ready?" I nodded. "Absolutely!" *T-Rex* would be back, in its originally designed mode.
>
> Bill Willard

It was heartbreaking to watch preliminary tests of *T-Rex*, intended to explore *Titanic*'s interior. Somehow, someone had made unauthorized changes to this mini-ROV, weighting it down to the point where it could not move. Its creators, Bill and Susan Willard, were angry and devastated, and given all they had invested in their invention, they had every right to be so.

Charles A. Haas

August 28, 1998

[Paraphrased] I was diving in *Nautile* on this day. We were more than 10 feet away from the D Deck door to get video of it. Though we never touched the door, it suddenly broke free and fell down. It is clearly visible on the video. I think that the hinges of the door were very corroded and fragile. Just going back and forth close to the door created some currents which may have been just enough to finish breaking the hinges. The heavy door was open for 86 years.

PH Nargeolet

There is a sudden commotion as everybody is suddenly alerted; a lift bag has surfaced close to the *Nadir*. From about a hundred meters we watch them recover the bag which contains the D-deck door.

Claes-Göran Wetterholm

In the days before the storm I spent a good bit of time going up to the bridge to track the progress of the hurricane. It was a bit disturbing to see the projected path going directly over the dive site. I was also in touch with Tim Foecke back at NIST who was thanking his lucky stars that I was on the ship and not him. Given we were down to one engine and the new one could not be brought onboard, many folks were rather tense. If we lost the one working engine, we all knew the ship would turn away from the waves and would likely roll.

Tim Weihs

I spent a lot of time on the Helicopter Deck. On this deck was the big camera, the *Magellan*. I noted in my diary on the 28th August the wonder of this enormous camera. I had never seen anything like it before. The previous night the little ROV, the *T-Rex* was tested.

Claes-Göran Wetterholm

August 29, 1998

It's been a perfect day of warm sun and sparkling seas – despite the unsettling news that Hurricane Bonnie may be headed directly for us. The forecasts, in fact, are looking pretty grim, with high winds and seas expected as early as tomorrow afternoon. But today, with nothing on the horizon but blue skies, it's almost impossible to imagine that the weather could turn so dangerous so fast.

Susan Wels

On one of the last dives, our ROV "pirate pilots" found a black valise handbag and lifted it by robotic arm and tucked it outside of the camera's point of view. The ROV was placed on deck, the black valise was serendipitously escorted to a secret, secret hiding place. [Immediately, this "piracy" was] reported to George. The ship-to-ship transfer of personnel to put the right people in the right ship was postponed as the seas grew in size and the wind was strong enough to breathe for you by just turning your head to the wind and opening your mouth.

Saul Rouda

Thus the legend of "The Whoopsie Bag" began. In the early morning hours, one of the ROV pilots did grab a valise in the debris field. There were many stories — one was that the pilot went to film the valise, and the robotic arm malfunctioned. Regardless, it had been made clear that there was to be no recovery of any item without approval from George or PH. Someone up late saw the bag, and informed George, who immediately came over with PH in a Zodiac. They immediately went into a closed-door session with the involved parties. One of the conservationists arrived with an artifact bin, in which to place the valise. Several team members gathered on the helicopter deck, sharing the news. "So, he just accidentally picked it up?" said one. "Whoops!" came the sarcastic response. So from that moment on, the valise was referred to as "The Whoopsie Bag."

> . . . they were not able to identify the culprits, so the search for intel expanded into the video recording studios which had tapes of the whole sordid affair.
> Before the search took place, one of the ROV gang flashed in and said they were searching for evidence in the videotapes, and could we help hide any relevant cassettes – just lose them for a while. I guess with the impending sense of doom from the storm at sea and our thinly veiled love of pirates, we did manage to "misplace" the tapes.
>
> Saul Rouda

The conservationist exited the office, and with help was carrying the artifact bin with the bag inside. A while later, an unhappy George left the room in a quick stride, followed by PH. But the issue was resolved. Within the hour, Jack Eaton joined the others on the helicopter deck having just awakened from his afternoon nap. Upon hearing the story, Jack commented, "Well, that sucks."

> Charlie and I took the tray on which the bag had been placed to the edge of the deck. In the tray was seawater and a quantity of what appeared to be rust-colored particles. Also in it was a broken clamping mechanism, in many pieces. I touched it, wondering who had been the person who had secured the bag during the trip. We returned the particles to the sea. I still wonder to this day who the bag belonged to, and what story it could have told us.
>
> Bill Willard

Later in the evening, news of the hurricane on its way became paramount in everyone's mind.

> The seas were rising and any "sailor's delight" on the moon was gone. Everybody wanted to pull up the ROVs, pack up and get out of Dodge to increase our chances of surviving the Hurricane Hell they said awaited us. Even the superhero ROV guys were ready to break camp. Our captains recommended that we stop work and run for it and travel some 100+ miles away from the center. Nice thought.
>
> Saul Rouda

In the morning, the decision was made to end the expedition early. The storm was now heading directly toward the team.

~ 10 ~

The Flyover

One of the most watched *Magellan* dives was the first flyover during the second half of the expedition. There were no requests for special shots for a television show, just historians completely glued to images of the ship. Occasionally, one person would make a drink run — a quick trip up to the galley to grab a handful of colas and whatever snacks that one could find. "I'm going for a drink, does anyone else want one?" was never asked, they would automatically bring an armload. The conversation during the flyover covered all topics — "What are we looking at? What happened at that location? What did it look like it 1996?"

The small makeshift amphitheater offered excellent views on the large monitor showing the feed.

> *Magellan* allowed us to see the metal rungs that had served as the lookouts' steps and handholds inside the mast leading to the crow's nest, and the cable emerging out of the crow's nest hatch, possibly the cord for the telephone that had warned *Titanic*'s bridge of the iceberg. The scant remains of the crow's nest were wrapped around the mast, dangling from the port side where the bracket holding the crow's nest remained.

Our Story

As our exploration with *Magellan* continued, I found myself comparing its images with what I remembered from my 1996 dive to the wreck. The forward edge of the bridge seemed to have slumped. At my request, we examined two circular controls on the inside edge of the bridge wing's window that had puzzled me in 1996. We now saw they were labeled 'speed control' and 'time control,' and were *Titanic*'s fog whistle controls. We spotted a brass knob at the outboard top edge of the bridge window, function unknown.

We returned to the bridge for our tour of *Titanic*, examining the crow's nest, the remains of the wheelhouse, the starboard bridge wing, the after wall of the grand staircase (looking for signs of a postulated pipe organ installation there), the gymnasium, the forward expansion joint, the forward starboard lifeboat davit and Captain Smith's quarters.

The deterioration since 1996 was striking. Now there were large holes in the starboard boat deck. The gymnasium roof had collapsed, and the top of the wall above the gymnasium's windows was gone. There was evidence of deck buckling. David Livingstone said decking over the reciprocating engines now was gone. The cranes forward of the superstructure were no longer parallel with the superstructure's face, leaning at about a 10° to 15° angle. There were holes above the windows at the forward end of A Deck, and in the decking outside the gymnasium, allowing three-foot by six-foot beams to show through.

PH described *Robin's* penetrations into the ship's interior. He said that C Deck had largely collapsed; where it hadn't failed, the columns of the deck were bending. The elevators' doors were gone, but one wall of B Deck forward of the elevators still had its wood paneling. Yet the interior walls of B Deck cabins were gone. He had not seen the purser's office.

Charles A. Haas

Several of us, perhaps six or seven of us, were sitting in front of the monitors on *OV* as the *Magellan* was doing a fly-by run. We were all watching in silent reverence, on the edge of our seats, focused on the haunting images before us. Jack Eaton, in pure Jack Eaton style, broke the silence, "William, I have a request." He was speaking to me. "Yes sir, Mr. Eaton – what can I do for you?" After a moment, Jack answered, "William, when you get your little robot down there, will you do something for me? Will you take a stroll with it along the Promenade Deck? It's been so long since anyone has been there." No one spoke. I stood there trying to say something past the lump in my throat, and eventually was able to say, "That's a promise, Mr. Eaton, if we ever get the chance."

Bill Willard

~ 11 ~

Pieces of the Past

Artifacts were recovered on several dives. Each piece tells a small part of the often-told story. Some artifacts tell of the forces that tore the ship into sections. Some artifacts show the grace and elegance from a dining table or kitchen. Others belonged to a passenger or crewman, and though we may not know the name of the person who wore a particular hat, or garment — that piece was a witness to history and has a story to tell.

August 21, 1998

PH climbs out, dressed in white coveralls. It was a frustrating dive. He and the two pilots – Patrick Cheilan and Xavier Placaud – were not able to bring up the door, but they did succeed in recovering a round electric light switch from the *Titanic*'s wireless room where, 86 years ago, John Phillips and Harold Bride transmitted the *Titanic*'s last urgent calls for help.

PH immediately gives the switch, in a water-filled plastic bag, to the expedition's conservators, Marielle Boucharat and Olivier Berger of France's LP3 Conservation. Marielle places it in a foam-lined basin and keeps it continuously covered with fresh water while they make their first assessment.

Although the light switch looks like it is painted white and red, it is actually made almost entirely of white porcelain. The red coloration, Marielle tells me, is merely the stain of iron oxidation. Three wires are still attached to the back surface of the switch, and the manufacturer's stamp is still visible: "LEKTRIK Patent Trademark," by appointment to His Majesty the King.

Susan Wels

August 22, 1998

Every object recovered from the *Titanic*'s wreck site — from the 25-foot-wide, 13-foot-high Big Piece to the fist-sized electrical switch from the *Titanic*'s Marconi room — presents a unique challenge for the conservation experts on the 1998 Titanic Expedition.

"Artifacts vary enormously in size, composition and state of preservation, and we take all those factors into consideration before we treat them," says Olivier Berger, one of two conservation experts aboard the *Nadir*. The Big Piece, for example, is made not just of steel plates but also glass, pieces of textile, bits of mineralized wood, lead pipes, copper-alloy porthole fittings, and rubber gaskets for the windows. There is even a fragment of a gold-rimmed porcelain plate that somehow became cemented to the hull section's outer wall.

Now that The Big Piece has arrived in Boston, conservators Stephane Pennec and Martine Plantec of France's LP3 Conservation will determine the best treatment given its complex composition. The first thing they'll do, however, will be to keep it in a wet environment to stabilize it while they plan further measures. Stabilization is essential before any conservation treatment, and it begins the moment the artifacts are brought up from the sea. "Our job on this ship," adds the *Nadir's* second conservator, Marielle Boucharat, "is to document the objects and keep them stabilized for travel to the lab." That means, most importantly keeping the artifacts moist. If they dry out, their inner layers could pull apart, potentially destroying the objects. Many recovered artifacts are small enough to be easily immersed in fresh-water baths in the *Nadir's* conservation lab. Because of its enormous size, however, The Big Piece required some creative thinking.

Initially, once The Big Piece had been winched aboard the *Abeille Supporter,* Olivier tried hosing down the huge metal hull plate, but the wind started drying areas almost immediately after they'd been moistened. Next, he tried covering The Big Piece with a plastic tarp to protect it from the wind, but the weight of the plastic threatened to compress the rusticles on the metal plate. Finally, he came up with an improvised sprinkler system that kept The Big Piece moist until it got to port.

"It took a few tries," he said, "but we finally got it right."

Documentation is the other responsibility of the *Nadir's* conservation team. All *Titanic* artifacts – even tiny pieces of metal that fell from The Big Piece – are thoroughly inventoried, measured, numbered, photographed and described as soon as they're in the conservators' hands, and the written records are entered into a computer database.

Olivier and Marielle also record the number of the submarine dive during which the artifact was retrieved, so that scientists and historians will be able to identify the area of the wreck site that it came from. And they take special note of any related objects that were recovered.

"The *Titanic* is not like an ordinary archaeological site," Olivier says. "We know what happened to the ship. But we are very careful with the artifacts so that we can add to that knowledge and help the world remember."

Susan Wels

August 23, 1998

There was quite a bit of excitement on the *Nadir* about a textile bag which the *Nautile* crew brought up with them from the *Titanic*'s debris field at the end of the dive. Perhaps it had belonged to a third-class passenger.

> We took a trip over to *Nadir* to view several recovered artifacts which, although their retrieval was not the expedition's main focus, nevertheless were fascinating. One was a grey burlap bag, about 8 inches wide by 12 inches deep, with metal grommets at one end tied together. Obviously handmade, it had a hole that had been neatly patched with white thread. Jack thought perhaps it was a ditty bag, or a handmade moneybag. (After the expedition, during the conservation process, it was found to contain a caulking compound.) We also examined a light switch from the wireless cabin and a gold-tipped plastic or celluloid cigarette holder.
>
> Another fascinating object was an entire panel of light switches from the refrigerators near *Titanic*'s galleys, each switch labeled with a celluloid tag. The D-Deck gangway door was nothing short of remarkable and magnificent. Among the other finds we saw were several tongue-and-dart pattern plates in a pattern known as *Wisteria*, a beautiful creamer decorated with hand-painted flowers, a teapot, an unusual hourglass that timed only single minutes, the remains of an electric clock from the wireless cabin, and a large pitcher with flowers on it, likely a passenger's possession and not part of the ship's equipment.

Charles A. Haas

Our Story

August 28, 1998

This afternoon, another piece of the *Titanic* rose out of the sea. It is the port-side gangway door from the D-Deck – well-built and well-preserved, despite 86 years on the ocean bottom. At 3:10 P.M. today, a lift bag filled with lighter-than-water diesel fuel floated it up from the seabed, where it had been lying close by the bow section of the wreck. It was then carefully winched onto the *Nadir*'s fantail and placed in a protective bath until it can be brought for treatment to the conservation lab in France.

It is a beautiful door, watertight and strongly built. Its eight brass locking mechanisms still shine, and a white rubber gasket still rims its inner surface. Two elegant windows, which can be raised and lowered, are still remarkably intact. The door was probably designed to be a first-class entranceway, since it opened onto the first-class reception area on D-Deck.

During the *Titanic*'s brief life, though, the door was never used for that intended purpose. First-class passengers in Southampton boarded through a gangway door on B-Deck, and those in Cherbourg and Queenstown came aboard on the *Titanic*'s starboard side. Nevertheless, it may be that crewmembers or passengers did attempt to use this D-Deck door—pitiably, not as an entrance, but as an exit, as the ship was sinking on the night of April 14, 1912. If so, this portside door may be linked to the last brave acts of seven deck hands on the night that the *Titanic* sank.

Susan Wels

Later in the evening I went over to the *Nadir* to view the recovered door. "What a beauty," George says. It's one of the entrance doors on the starboard side, probably First Class.

Claes-Göran Wetterholm

August 29, 1998

At 6:30 this evening, when the sub returned to *Nadir*, it brought with it some delicate artifacts from the *Titanic*'s debris field. The most remarkable was a golden chandelier from one of the first-class public rooms – its ornamentation and gilding virtually unmarred by more than eight decades on the ocean floor. The chandelier is just one of many ordinary and extraordinary objects that *Nautile* has recovered during this summer's expedition.

One of the smallest objects that *Nautile* recovered – and one that is particularly intriguing to historians – is a simple coat hook from one of *Titanic*'s staterooms. It's of interest because before the *Titanic* left Southampton, her designer, Thomas Andrews, concerned himself with the tiniest detail of the ship's décor – even noting that in the future, Harland and Wolff should consider "reducing the number of screws in stateroom coat hooks."

Now for the first time, we know that there were exactly four screws in the *Titanic*'s coat hooks.

Other notable artifacts include a light-switch control panel from the *Titanic*'s kitchen area, with individual switches labeled "Mutton," "Portable Mutton," "Fish," "Portable Fish," "Thawing Room" and "Poultry and Game;" a skylight from the first-class elevator shaft; and, almost miraculously, a tiny 60-second hourglass that remained unbroken despite the force and violence of the liner's sinking.

Susan Wels

There is also some kind of a board with buttons and my guess is that it comes from one of the cooling rooms where the ship broke and everything was scattered around – including myriads of fragments of cork that covered the sea. Some of these cork fragments were noted when Lifeboat A was discovered on the 13th May 1912, and three dead men were found. They all had cork in their mouths and there were rumors of them starving to death and desperately trying to eat their life jackets.

Claes-Göran Wetterholm

Perhaps the most intriguing of all is a round dial, with a spindle and spring mechanism, that may well have been the electric wall clock from the *Titanic*'s wireless room.

Susan Wels

We also looked at a different kind of china; we assumed it was from Third Class or crew. It's a completely new kind of china I haven't seen before but later I've thought of it as maybe coming from a passenger.

Claes-Göran Wetterholm

~ 12 ~

Hurricane Bonnie

> Although Hurricane Bonnie has been roaring through the North Atlantic, her eye, and her full fury, is passing north of our position. So at least for now, we're staying here, on the *Titanic* site. Expedition leaders knew we'd be in for some wild weather, but they were also confident that our ships could take it. Whether the passengers can handle it or not is another story.
>
> Susan Wels

WHEN EVERYONE MET in the galley early in the morning on August 30, people were focused, not their usual jovial selves. Terseness replaced conversation as the team filled their coffee cups, some fixing a small breakfast. The decision makers were upstairs in the bridge office. Finally, one of the crew came in and began to explain the current situation. Hurricane Bonnie, on course to miss the expedition site the day before, had taken a turn in the night and was heading directly toward the site. "We should be fine," he added. "We'll move out of the path, and ride edge of it." Anxious faces stared at him and said little. Fear does that to a person at times. The crewman, Jimmie James, smiled and tried to allay everyone's fears. "Oh, I've been through a'many of these – we'll be fine and better – you'll have a great story to tell when you get home!"

Outside the swells had increased from the day before, and were still increasing, based on what Jimmie James had said. The skies ahead were dark gray. From the upper decks, those who had gone out to take a look could see wave after wave queued, on their way to our ship. Off to port *Nadir* was several hundred yards away. Another crewman was tying ropes from the forward section door to the aft section door. He explained the rope was "to hold on to — this is going to get rough before it's over. Don't come outside." His eyes cemented the importance of his last sentence.

The decision was made to vacate the site. The rest of the planned work would have to wait until the next expedition. The task at hand was to prepare for the approaching storm. First priority was to transport people to their appropriate ships for leaving the site. *OV* was heading to Boston, and *Nadir* was heading in a different direction.

> On the 30th the weather began to deteriorate and the sea was building up. At the same time, there was a transfer of people and luggage between the *Nadir* and the *Ocean Voyager*. I couldn't understand why this took place now; earlier the sea was flat calm and transfers could have been undertaken without any problem at all, but now this was extremely risky.
>
> Claes-Göran Wetterholm

> We are riding the hurricane's tail. As experiences go, it's a wild one. Earthquakes usually rock you for a few minutes at most, but these hurricane-force gales and seas have been slamming us senseless hour after hour, all day and all night, flinging us into walls, hurling our belongings on the floor and flooding the ship's fantail chin-deep in churning water. And we're not even in the direct path of the storm.
>
> Susan Wels

Before the storm hit in earnest, some of us ventured up to the bow and watched as the ship dove down into a trough and then climbed up the next wave. The spray and the noise were intense.

Tim Weihs

By this time the seas were eight feet, feeling like 30 feet, and we still had to transfer people ship to ship on small rubber Zodiacs. In normal seas the transfer takes place the moment the two ships reach an equilibrium; they both stop for a second. Well, with 8-foot seas there is 16 feet of vertical wave action. We go from safe ship, through the air to little rubber boat back onto big safe ship. First we pass our camera gear, then with experienced boatmen calling the shots, and most importantly calling the timing, we must jump . . . and then be caught and gathered into the Zodiac boat. Then grabbing hands land you again on a big safe ship.

Saul Rouda

Hurricane Bonnie arrived in full fury from the west. The experience was frightening, not because of the wind or the waves, but from the realization that if the remaining propulsion failed, we likely would turn broadside to the wind and capsize in minutes, especially with all the extra weight of the vans installed on the upper deck. And there was the realization that the immersion suits wouldn't be much help. I saw the tops of waves being blown off by the wind, and water coming up to the forward end of the work deck, surging between our cabin and the dining room. Taking pictures outside – indeed, *walking* outside - was dangerous.

Charles A. Haas

Our Story

As the seas became rough just prior to Hurricane Bonnie, Dr. Cullimore did one of the last transfers from the *Ocean Voyager* to the *Nadir*. Given only minutes to pack and get on deck, it was a rough ride over to our new home that was going back to St. John's. Once onboard, again it became an issue as to where to put me. I was given the choice to stay with Dr. Cullimore or a crewmember . . . so Dr. Cullimore it was! We were put at the very bottom of the ship, which was fine by me.

Lori Johnston

Late in the morning we did these transfers . . . mostly smooth but George got stuck in the classic position of being between two ships, and not on either one. He held on to a rope from *Ocean Voyager* and dangled as the ship's side sunk beneath the waves as did George. He emerged holding tight and was hoisted aboard.

Saul Rouda

For me personally, it's been a challenge – especially since this morning I had to transfer with all my gear from the *Nadir* to the *Ocean Voyager*, the ship that will bring me back to Boston. That meant another dreaded Zodiac ride – this time in seas that were 10 feet high and climbing. I had to crouch far back in the inflated motorboat so some mongo wave wouldn't sweep me out to sea. The real thrill, though, came when the Zodiac negotiated the cliff-like waves and arrived at the *Ocean Voyager*. Naively, I had been expecting to step out of the Zodiac onto the *OV*'s fantail. If I had to come aboard on the ship's side, I imagined the crew would at least unfurl a sturdy rope ladder for me to climb. No such luck. As Zodiac drivers Julien Nargeolet and Max Salmhofer (true heroes, in my book) struggled to maneuver close to the side of the *OV*, rising and dropping precipitously on the huge swells, the ship's crew tossed out a narrow knotted rope for me to climb.

Now, I could never climb a rope as a kid in P.E. class. Why in the world somebody thought I could do it with wet hands, in the middle of the North Atlantic, with a hurricane coming on, is beyond me. But, as I've learned after getting in and out of many Zodiacs, the rule of thumb is "jump or die."

So as the boat lifted on a wave, I grabbed the rope and tried to pull myself up the six or seven feet to the ship's rail. No chance. My rubber boots couldn't get traction on the *OV's* slippery hull. And then the Zodiac pulled away from me on a wave. For a few seconds, I was dangling from the side of the *OV* like Kate Winslet hanging from the stern of the *Titanic*. Only now that I know what it feels like, I know she didn't look anywhere nearly scared enough. Fortunately, on the next upswell, Julien gave me a push from underneath, and up on deck, Tom Dettweiler managed to haul me aboard the *OV* by my wrist.

Half an hour later, in even worse seas, George Tulloch tried to come aboard by the same rope and got brutally slammed between the Zodiac and the ship. Luckily, he managed to hold on and was pulled aboard – but if it was as scary for him as it was for those of us watching, it'll be a long time before he tries to board a ship like that again.

Susan Wels

Hurricane Bonnie virtually flew up the east coast of the United States, and the expedition ships had to leave for safety's sake. A Zodiac brought team members to the *Ocean Voyager*. As it came alongside *OV*, George reached for the rope leading to the deck and to safety. As he did so, a huge wave hit him and pulled him out of the Zodiac; he dangled from *OV's* rope up to his waist in roiling water until PH and Julien Nargeolet and the *OV* hauled him back in. Our hearts were in our mouths.

Charles A. Haas

Lunch was hastily made as the swells increased in height. In layman's terms, an 8-foot wave is measured from the rest position, the flat position of the wave. Each wave has a crest and a trough — so from the highest point to the lowest point is 16 feet. The storm started slowly — with a long time between peaks — and increased as it grew nearer. By lunchtime, the ship would go from a 30-degree angle uphill to a 30-degree angle downhill, repeatedly. Lunch was sandwiches, chips, and for once, no rice.

Braver team members ventured out on the bow to experience the storm. The Ship's Master allowed it early on, but eventually said a resolute "NO" as the intensity increased. People began to disperse slowly; most went to their cabins and nearly everyone put their survival "Gumby" suit within reach. Many opened their room portholes to see the ominous gray overhead and the massive, relentless wall after wall of water.

My cabin was just across from the main restroom. Certain bodily functions do not stop during a hurricane. In mid-afternoon, one at a time, several men had gone in and exited the restroom with soaking wet pants. It was not toilet or sink water, they had wet themselves. One usually does not want to laugh at the plight of another, but this was not one of those times. "What in the world happened?" I queried. I will paraphrase the common answer: "It is not easy to control the fire hose when I had to use both arms to grab and hold on to keep from falling." So I asked, "Why didn't you sit down like the ladies?" The facial expressions were priceless!

Bill Willard

After all the excitement this morning, I went to bed. Normally, that would be a calming experience. But as the winds and seas continued to get wilder, even lying on a bunk became almost unimaginably stressful. I was in a top berth, inside the bow, and as hurricane winds of up to 70 miles an hour shrieked over the water, I heard huge waves pounding the ship with body blows that boomed through the steel hull. Some of the waves generated so much G-force that they levitated me right off the mattress, and I had to brace myself against the top of a locker to keep from flying off. One enormous wave flung me off the ladder as I was climbing to my bed and hurled me bodily across the room.

Susan Wels

The people in the lower cabins, such as those on E-Deck, heard terrifying noises.

    Eventually, things were not only sliding back and forth but crashing all over the place. The waves were getting higher.

Cindy Tulloch

I took photos of the bridge's wind speed indicator, showing 50 knots with gusts to 60, which equaled 70 land miles-per-hour, while seeing waves taller than *Ocean Voyager*. Finishing my storm picture-taking session, I was pitched down the stairs from the bridge, painfully wrenching my arm. At full speed on our one functioning screw, we moved less than 20 miles from the *Titanic* site.

Charles A. Haas

    As afternoon turned into evening, the slope of the deck increased, as did the wave heights. There was no dinner. Those who wanted something could venture into the galley. The refrigerator was stocked with drinks, there were snacks available, but the galley was empty. On the bridge, all hands were at the ready, with the Ship's Master steering directly into the repetitive walls of water that were battering the ship. It was difficult seeing the top peak of the approaching wave without being up at the forward window. The slope of the ship neared 40 degrees, and now the waves were hitting faster. Most people stayed in their cabins.

> I had a top bunk and spent hours with my head hitting a wall as we went up the wave, then my feet hitting the opposite wall as we rode down the wave. It was around 8 P.M. when we began to ride the most intense waves. As we rode down a wave, the ship's anchor would move away from the ship. On the quick upturn, the anchor would snap back into place with a loud collision, a metal-on-metal clanging. At first many feared the anchor might puncture the hull, but we tried not to think of that. The anchor hit the hull for hours. Very few people slept. It was approximately 3:30 A.M. ship's time when the clanging finally ceased, bringing relief to many of us. Within the next hour, the slope of the ship returned to the 30-degree slope, then less. Many fell asleep from emotional exhaustion around this time. The waves had reached a maximum height, as told to us by the bridge crew, of 42 feet. In simple terms, we rode 84-foot waves (from peak to trough) for six to eight long hours. All total, we survived a 24-hour moment in time that many of us will never forget. As Jimmy James so eloquently put it – we had a great story to tell when we got home.
>
> Bill Willard

> That night I stayed up on the bridge and watched as we rode out the storm. I figured with a life jacket on and being up on the bridge, I'd stand a decent chance of getting out of the ship and surviving a roll. The warm water temperature also put me at ease, to some

degree. I knew I could last for a while in these water temperatures.

As we rode out the worst of the storm, I have clear images of us being down in a trough and looking up at the crest of the next wave. The oncoming crests were well above the 30-ft high bridge on which I stood, letting us know how big the waves were. That night I came to appreciate the immense power that the sea can bring to bear.

Tim Weihs

At 8:00 P.M., the really big seas started coming on – ranks of 30-foot waves, one after another, and gusts of over 80 miles an hour. As a somewhat rash antidote for seasickness, some *OV* passengers have gathered outside the bridge tonight to watch the full force of the weather. The ship is facing straight into the waves, and they watch as each huge wall of black water rises above them, curls and crashes on the bow. None of them have seen anything like this before. In weather this big, you begin to feel that a 200-foot ship offers only a little more protection than a tin can. And for most people who spend their lives on land, 30- to 50-foot waves, in the middle of the North Atlantic, are in a category of danger all their own. "To understand waves that big, you need a geology degree," says Discovery Channel cameraman Mark Knobil, "because those aren't waves, they're mountains." Fortunately, after tonight we'll be saying goodbye to Bonnie.

Susan Wels

From after noon until nine the next morning I stayed in my bunk, trying to keep in the bunk and not be thrown out of it. It was a desperate fight. My cabin was in the bow, so I had a roller coaster ride (which I hate!) for 20 hours. I heard later that the highest wave was 13 meters but most of them "only" 10 meters, which meant crushing down in a valley of water but seconds later going upwards, and this hell never seemed to end. Suddenly I realized there was probably more water in the cabin than outside. I had to force myself to leave the bunk and try to reach the porthole which was only partly closed. I overcame myself and finally managed to close it. The sound of "The Big Swell" is terrific and that combined with the constant banging of one of the anchors made any attempt to sleep in vain. (I remember being in the bunk and saying to myself out loud, "What the hell am I doing here!?!")

Claes-Göran Wetterholm

Once the hurricane took over, we [Dr. Cullimore and I] both basically stayed in our bunks trying not to be smashed and bruised. At some point I had decided to go up to the bridge to see what the heck was going on. It took a ridiculous amount of time just to get up to the bridge as the movement of boat was so incredible. Once at the bridge I realized I should have brought my camera, but my thinking at the time was indeed hazy at best. It was just walls of water that surrounded us, ludicrously large walls of water that when they broke over the vessel, the inside was black as night. It was really surreal. After that scene, I remember turning around and thinking I might just need a nap!

Lori Johnston

I was up on the bridge at one point when the waves came right over the bow and the two lower decks and right up to the bridge's window. At that point, I was really afraid. I knew that we were just in the beginning of the storm. I tried to picture what the middle and end would be like. Making it to the end was what terrified me.

Cindy Tulloch

All were at general quarters and the film [crews] and ROV gangs were free to "roam about the ship." The ship's doctor brought out his emergency medication to fight off fear and seasickness. He then handed us each a little white paper pill cup and asked us how many Valium we would like. Yes, Valium could save the day. I asked for some Valium for me and some for my wife who was anxious and worried back at home in California. We were on a "dry ship" of British register . . . so it was back to the paper bag trick; the liquor came out to steady our nerves. The storm was getting crazier so we took our grog (any drink in a storm is a "grog") and headed up to the bridge control center. Our captain valiantly and confidently kept the ship on course, heading into a three-way sea.

A one-way sea has one set of waves mostly in parallel. If there are two storms causing waves, there is a cross sea, where the energy of each wave matches and builds on its counter-wave from the other direction. With three storms each generating waves, we have the worst of three possible worlds. The waves are forced up high and the ship rides each wave, rides unto another wave from another point of the compass. So it's up one, up two and then up three and then the bow is pointing upwards and stays there for a brief moment, enough time for us to scurry a move about the ship.

[We left] the comfort of the bridge and went outside in 60-mile-an-hour wind. That is not just windy, that's "push you over and fall down" windy, that almost lifts one into the air and away. Adrenalized, in a fight/flight mood, [we were] yelling the name of the hurricane,

"BONNIE!" in defiance outside, sitting with our legs dangling over the catwalk in front of the bridge, where we were totally alive. Then the ship plunged 60 feet into a waiting piece of the sea, dug itself out and her bow cut into the deep green sea wall. The water leapt only to be picked up by the speeding wind, traveling as a foamy water-filled cloud. This hit us hard. Luckily there was a warm rain a'falling. It was like a five gallon bucket being thrown at you hard every ten seconds. After having enough of a purgatory beating we retired to our bunks, planning a way to escape if suddenly there was water in our cabin. Maybe that is why Mark, my cabin mate took the upper bunk. And I thought he was being nice and that he didn't want to be stepped on in the middle of the night when pee must out itself.

Saul Rouda

I am lucky to be alive. All of us on the *OV* are lucky to be alive. We donned our survival suits. Put the suggested orange and bottle of water in our survival suit pockets. And then hunkered down in our berths for the next 18 hours or so while being tossed about like rag dolls.

Charlene Haislip

Little did the team realize that the situation was even more serious than they imagined. During the night, the ship was running on one engine. One engine was just enough to enable the ship to turn into the waves.

The *OV* did not have enough engine power to get out of Bonnie's way. Our only option was to try to hit Bonnie squarely head-on so that our bow would neatly slice perpendicularly through the oncoming waves instead of being hit broadside and then capsizing. Our expedition physician, Dr. Budman, passed out Styrofoam cups filled with sedatives, warning people not to take "too many." Sedatives would help our nausea somewhat, but if we actually had to use our survival suit, we'd be [in deep trouble]. My berth was on the lowest deck, right next to the engine room. I curled up on the top bunk because we were taking on bilge water and that water covered the bottom bunk. I spent the entire 18 hours on that top bunk, projectile vomiting, trying to ignore the sloshing "waves" of the water in the room below me.

It wasn't until very recently that I found out that we all did come within a hair's breadth of perishing. Seems our one working engine did stop running for a period of eight minutes. Just like a scene from some disaster movie, maybe even a *Titanic* disaster movie, *OV's* engine crew strapped themselves to poles so they could try to work on the engine without being thrown about the room. Lucky for us they were successful. They got the engine going again before a wave hit us broadside.

Charlene Haislip

My wish we should be spared from the hurricane was in vain. Charlene from Discovery was gone for three days and when she finally appeared several of us began to worry seriously. She had hardly eaten and I said to her, "Charlene, it's a tremendous way of dieting – but it's not fun!"

Claes-Göran Wetterholm

Memories of the storm are vivid. Jack and I watched cascades of "stuff" falling off our cabin's desk, shelves and window alcove. Concerned that I could be pitched out of my upper bunk by the ship's motion, I took all the bedding and moved down to the floor, Jack watching in amusement. Sleeping did not come easily, as the heavy rocking was accompanied by pounding, metal-on-metal sounds, and, remarkably, some noisy people outside our cabin until 1 A.M. – a storm party, perhaps – or people preparing to abandon ship?

  Later that night I awoke to find Jack standing over me. My mattress and I had been sliding back and forth across the room's entire length, and now were blocking the entrance to our cabin's bathroom. We learned later that the storm's maximum wind was clocked at 85 knots (98 mph) at about 11 P.M., with waves about 15 meters (50 feet) high.

Charles A. Haas

My memories are a patchwork of moments I will never forget. The hurricane is the story I retell most often. The power of nature in the churning sea was humbling and amazing at the same time. I was one of the crazy ones that sat on the bow and rode up and down the waves screaming, "Come and get us!" It was a powerful science lesson in buoyancy as we would watch the bow dive under the waves to reappear a moment later as it jerked us straight up to ride the next wave. No amount of those little white pills the doctor was freely handing out could douse the exhilaration of that near-death experience. Months later I read *The Perfect Storm* and I was so glad I was ignorant of the phenomenon of rogue waves. Remember how we had to time our trips up the ship's ladders to coincide with wave riding? When weather people on TV say ". . . and the hurricane will go harmlessly out to sea" I think to myself, "Well, not if you are in a tiny boat in the unpredictable path of the storm!"

Stephanie Ratcliffe

The next morning we heard what happened [during the night]. According to the crewmen, the ship tilted almost to the point of capsizing. These crazed moments seem worthwhile if you survive. Adversity only makes us wiser not smarter.

Saul Rouda

Even to this day, I will not watch *A Perfect Storm*.

Bill Willard

And then there was this:

Dick was a sailor – a true Navy man. He didn't talk much about the experience of the hurricane. I doubt it bothered him.

Beverly Silloway

~ 13 ~

Homeward Bound

The *Abeille* arrived in Boston on the evening of August 20. A conservator would meet the ship and determine the best course of action to preserve The Big Piece. Across the street from the ship's docking berth, the Boston Titanic Exhibition was under way, and many guests had made reservations for the exhibit with high expectations of seeing The Big Piece and experiencing its significance.

> When we approached Boston on the evening on August 20th, The Big Piece had been covered. It was to be unveiled to the public on the next day. As we neared the dock, we met a boat with lots of photographers. "This is the press coming our way to get first photos of The Big Piece," we thought. It turned out that it was just a whale-watching boat.
> This was the first time the *Abeille Supporter* arrived in the US. Everyone was nervous about the customs formalities. When the customs officers came onboard, they searched the ship. They found the wine cellar and could not believe the significant amount of wine they discovered aboard the scientific vessel. They were convinced that there was some type of wine smuggling in progress.

However our captain convinced them that all the wine is for the consumption of the crew. The wine store was then sealed for the time the *Abeille* was in Boston.

Soon the rumor made rounds that Celine Dion would unveil The Big Piece. She was in town; her world tour had its premiere on August 21 in Boston. I was glad that there was no crossover with the movie and that it remained a rumor.

Since we did not know when exactly the *Abeille* would arrive in Boston, I had booked my flight for August 24; therefore, I stayed four more nights on the *Abeille* and enjoyed the city and the nearby exhibition. The removal of The Big Piece was scheduled for August 24th, and I was optimistic that I could see it being lifted to the shore because my flight was in the evening. The *Abeille* was moved to another berth where a huge crane would have better access. The crane arrived and prepared everything. However, the captain denied the removal of The Big Piece, due to the fact that the chartering fees for the *Abeille* were not yet paid. Things became hectic because RMST had to organize an express money transfer to France. Time passed. Everything was ready, all the lines were attached, and a test was made to lift it off, but only a few centimeters.

I started to fear that I could not stay long enough to see it being lifted off. RMST tried to talk the captain into giving the permission, but the captain had his instructions from the office. It was time for me to leave for the airport. I asked one of the police officers how long it would take to get to the airport. The officer understood how much I wanted to see the Piece lifted off the ship. He asked me for the latest time at which I could check in. He told me to stay to the last possible moment and he would drive me to the airport. He even offered to use the siren in case of a traffic jam. This gave me more time – but the confirmation of the

Our Story

received payment from France did not arrive. As promised, the officer drove me to the airport. There was not much traffic, no sirens were needed and I just made it to check in before it closed.

Günter Bäbler

Back out at sea, onboard *OV* and *Nadir*, the storm had passed, and the teams awakened to a brilliant sunny day, the ship on course headed westward toward Boston, while *Nadir* was headed to Canada. Many of the *OV* crew slept in after spending most of the night clinging to bunks and survival suits.

PH told us we were expected to reach Boston on the 5th September, and after the hurricane experience we all cheered up. Then I remembered what one of *Nautile*'s pilots told me during the 1994 expedition. *Nautile* had been somewhere in the Eastern hemisphere when a typhoon struck. There was no time to bring the *Nautile* up; all they could do was tow it behind the *Nadir*, hour after hour. "But what do you do [when that happens]?" I asked. "You wait and vomit," he answered.

Claes-Göran Wetterholm

Our near-death experience did not negatively impact either the expedition or the production. True, we hobbled back to Boston port. And from there, all of us hobbled back to our respective homes. Stardust finished all documentaries on time, on budget, and to much critical acclaim. In fact, the two-hour live telecast, *Titanic Live*, was Discovery's second highest rated program in its history (4.1 rating). And the documentary, *Titanic: Anatomy of a Disaster*, broke all records drawing a 4.5 rating, which put it at #1 at that time. However, the near-death experience of sending an entire boatload of people straight into the jaws of a hurricane rather than spending

$15,000 to fix that boat's engine is precisely the reason I quit production work after this *Titanic* series. This was "the" experience of my lifetime. I just wanted to make sure my lifetime lasted a little while longer.

Charlene Haislip

Dick [Silloway] hoped to dive [in *Nautile*]. He was a submariner early in his naval career. The hurricane wiped out his opportunity to dive.

Beverly Silloway

Late in the morning, several members of the team made their way out to the deck, grateful for the sun and calmer seas. Having feared the night before might be the final night of their lives, quite a few were grateful just to see a new day. They shared stories of their experiences during the storm, laughing and commiserating where appropriate. Someone noticed debris in the water. Small items, sometimes larger items, each was a reminder of the storm as it bobbed on the surface as the ship passed by. They identified fishing floats about the size of American footballs and bright orange, large parcels of rope, broken wood slats, and at one point, a basketball. They were close enough to it they could see the logo of the sporting goods company.

> Charlie Haas asked to get a photograph of the *T-Rex* with its creators. The ROV was brought out on the fantail. Charlie was joined by Claes-Gören, and they both took photos. Charlie used his photo in his [and his co-author Jack Eaton's] wonderful work *Titanic: A Journey Through Time*. I believe it to be the only photo taken of me during the entire expedition where I wasn't smiling. After being awake 28 consecutive hours, my scowl was the best I could do!

Bill Willard

On the 2nd September we had a big party in the dining saloon, and I drank a little too much. The next morning I wondered how anybody could eat anything whatsoever. The daily morning ritual was undertaken, "Burn the toast." The toaster was probably beyond retirement, and there was always the choice between "not toasted at all" and "coal."

Claes-Göran Wetterholm

As time passed, the ship moved farther from the last remnants of the hurricane. The seas calmed, and everyone began to relax and enjoy the ride once again.

Later in the morning we enjoyed the calmness on the helicopter deck. It was me, Cindy (Tulloch), Gary (Hines), Susan (Willard) and Bill (Willard), and we talked about the hurricane. I noted that the weather [had improved] and the waves [had calmed] as we distanced ourselves from that experience.

Claes-Göran Wetterholm

Meals onboard were generally wonderful, but every day there was rice, plain and boiled and tasteless – every day, and every meal (except breakfast.) By the expedition's end, I could no longer eat it. Some of us began thinking about finding other "uses" for rice, including a rice sculpture contest and perhaps even a "Rice Olympics" with all events based on rice. (Down below, we later discovered 500-pound bags of the stuff!)

Charles A. Haas

Tom Dettweiler called a meeting in the galley. All hands in, except for working crew. There had been an awful tragedy in Halifax.

OV was slightly southeast, heading westward, a day from Halifax at most. Swissair Flight 111 had crashed five miles offshore from Halifax, with no survivors. Tom informed the team that *Magellan* might be helpful, but he wanted the team to discuss offering assistance.

He said, [paraphrased] "All of you have regular jobs you have to get back to. However, *Magellan* is a remarkable ROV that can deploy immediately upon arrival, and we might be able to be of great assistance." Questions were asked, including "Were there any survivors?" And Tom, soft spoken and serious, replied, "It is highly unlikely." After a long silence, someone spoke [again paraphrased]. "How can we refuse if they need us?" Heads slowly began to nod, eyes began to look up to Tom, standing with his arms propped back on the counter. Tom asked, "Are we in agreement, then, to go if they need us?" All who answered said yes or nodded their heads. "I'll go tell them. I'll get back with you."

> In the morning I walked into the bridge. I overheard a strange conversation between our ship and the Canadian Coast Guard. It was difficult to follow but they asked us about the *Magellan*. A large plane had gone down. I came to understand more. It was a Swissair plane traveling from JFK to Geneva. It seemed we were the closest ship. I noted there must have been over 200 people on it. Finally I heard we were not needed, despite the *Magellan*. The Coast Guard seemed to have enough equipment to find the plane. Later I read that everything was smashed to smithereens. How strange to come from the world's most famous shipwreck and be so close to another horrible disaster.
>
> Claes-Göran Wetterholm

That evening, an event unfolded that, for a short while, brought measurable concern and fear to many. Gary Hines was walking outside as the darkness was settling in when he heard a woman

Our Story

scream. Quickly, he looked around and saw no one, but hastily made his way to the room where all the women were sitting and talking. He only counted six of the seven women onboard in the room. Susan Willard was missing. Gary quickly found Bill in the galley observing another Scrabble battle, and suggested he check their cabin for Susan.

> She was not there, nor was she in the restroom, galley, bridge office or anywhere else. Gary informed the captain what he heard, and that no one could locate Susan Willard. We feared she had fallen overboard. The captain immediately slowed engines and called the crew into action. The ship had travelled eight to ten minutes away from the approximate spot of the scream. I immediately went to the helicopter deck, from where I would have the greatest vantage looking on the ocean. Fear was escalating for everyone who went up there with me. Gary went back to the room with the women, and told them he had heard a woman's scream, and the fear was that someone had fallen overboard. He said at that point, "No one can find Susan Willard." Sitting in the room, in the corner, Susan leaned forward and said, "I'm right here . . . " Gary breathed a great sigh of relief, "Were you in here when I came in a few minutes ago?" One of the ladies replied, "She's been here the entire time." Gary immediately notified the bridge all were accounted for, and the course resumed. The small group on the helicopter deck was very relieved and thankful.
>
> It just so happened that in the galley a small group had been watching a slasher film, and just as Gary walked by the porthole, the slasher claimed a victim – that was the scream he had heard. It was another story to tell, another emotional moment not on the expedition agenda. I was thankful that Gary investigated as

he did – his quick actions could have saved a life if the situation had been different.

Bill Willard

Bonnie wasn't the end of our meteorological adventures. As we recovered from that storm, the fringes of another major hurricane, Danielle, hit us on the way to Boston. It was followed by a day of dense fog.

Charles A. Haas

On the morning of September 4, all team members were on deck. *OV* was just a few hours from Boston — almost home.

September 4th is the last morning aboard. "No rice!" someone had scribbled on the white board in the dining saloon. The cook must have been sponsored by either a rice or a salt company."

Claes-Göran Wetterholm

From the bridge, the radar indicated possibly a hundred vessels in and around the *OV's* path. As *OV* neared the vessels, they were identified as Boston fishing boats with their crews busily checking traps, bringing in their catches, moving to their own locations among the fertile fishing areas. *OV's* course was computerized, and the Ship's Master did not slow nor veer from the course. This upset quite a few of the crews on these nearby boats. Members of our team lined the rails, from the helicopter deck down, on both sides of the ship, smiling and waving as *OV* passed. Those fishing crews returned a different "salute" and shouted things that couldn't be heard, but it was obvious they were not happy comments. After the *OV* passed through the armada, Boston's skyline became visible.

Everyone made their way to the helicopter deck and began taking their last expedition photos. Charlie Haas owned a burgee [a flag used by ships for signals or identification] that flew aboard the motor vessel *Britannic* (1930-1960), the last White Star ship in service. The Ship's Master allowed the burgee to fly from *OV* as this was the last hours of the trip home. A group photo was taken, several of them as a matter of fact, and then individuals posed with each other. George and PH, George and PH and their wives, then Jack and Charlie, and so on.

In a short time, the pilot's ship arrived, and the pilot boarded *OV* to take her into the harbor. PH and George had a brief meeting with the team while everyone was together in one place. He reminded them there would be press — lots of press, and there would be television cameras, still cameras, voice recorders, and many, many reporters. He asked everyone to be positive and to enjoy the moment — they had earned it.

> We made a spectacular entrance into Boston Harbor, with my White Star Line commodore's burgee flying from the *Ocean Voyager's* mast in brilliant sunlight as we were escorted by all sorts of harbor craft. Jack and I were invited into the wheelhouse at the request of the pilot, who wanted to know more about the flag.
>
> Charles A. Haas

Two of the Stardust production assistants, Brandon Plonka and Steve Ulrich, both did a yeoman's amount of work. They took the two stewards under their wing. Those two young men worked very hard on the trip. In addition to taking our meal orders and serving us, they would bus our tables and clean the galley. They also would clean our rooms – sweeping them, mopping them as needed, and then cleaning the halls and other areas of the ship. The two stewards knew limited English, so Brandon and Stevie began to teach several "American" phrases to them. These phrases, we discovered later, had colorful metaphors in them. As a reporter went to interview Daniel, he stood proudly and looked the camera square in the lens, and spoke a phrase he had learned. PH was nearby, and hurried to the scene, and talked in French to Daniel. Daniel's face turned to horror when PH translated what he had said. After Daniel's explanation, PH graciously told the reporter of the practical joke played on the innocent steward. The reporter understood, and PH only said, "Has anyone seen Brandon and Stevie?" The reporters dispersed gradually, and several of the team prepared to depart for home.

Bill Willard

Goodbyes were said. Some team members walked over to tour the Boston Exhibit. The Big Piece was not in place yet; it was undergoing treatment in a nearby facility. Most of the team would not see it again for many years.

We were on dry land at last. I was surprised by the rubbery feeling in my legs after six weeks at sea. As we disembarked, we were surprised to find a delegation of Titanic International Society members there to greet us. It was with supreme regret that we could say little more than "Hello" and "Good-bye" to them, because we had a plane to catch.

There was an emotional, final round of good-byes . . . and then, only memories of the best damned expedition *Titanic* ever experienced, filled with new understandings and knowledge, worldwide public participation and fascinating artifacts.

But there also had been long days of intense, relentless work; supportive, collaborative friends and colleagues; elements of unexpected danger; and, ensuring this expedition's legacy to the future, two remarkable men: Paul Henri Nargeolet and George Tulloch.

Charles A. Haas

That evening, Discovery hosted a dinner for those who remained in Boston, in the upper room of a local restaurant. The food was splendid. Several people stood to share special comments, and everyone enjoyed each other's company for one final night. The next morning, most of the team departed at different times during the day, saying goodbye to their ocean home, *OV*, and spending the final night in a local hotel.

The old adage is certainly true: There's no place like home.

Reflections

The 1998 Titanic Expedition? It was a big adventure. I had never done anything like this!

Steven Biel

The end of the story became a bit more turbulent than I ever could dream. My sister, who lives in Boston, met me at the ship. She planned a trip to Maine where we would meet a friend of hers and go canoeing in the wilderness. In the evening there was a telephone call from Maine telling my sister that her friend had died of a heart attack and our trip was cancelled. Back home in Stockholm I scheduled an appointment with my doctor and had an MRI. During the whole expedition I had a severe headache and no medicines helped me. The doctor told me I was suffering from something he called carotid dissection. Minutes before I was rushed into the intensive care he said, "NO exercise! If you as much as do jogging your artery can burst!" It struck me then and there how strange life is: Had we gone canoeing in Maine it was a great possibility this would have happened. How strange, I thought to myself, that someone else's death had saved mine. But that's a different story . . . .

Claes-Göran Wetterholm

Our Story

Author's note: Considering all the "exercise" everyone did trying to maintain their balance or secure positions in their bunks or cabins during the hurricane, Claes-Göran was very lucky indeed.

> It is the portholes that I suspect I remember most [about The Big Piece]. There are four of them, two big ones and two small, as well as portions of a fifth and sixth. Three still have some of their window glass in place, and two have beautiful brass fittings that look practically new. The brass is still engraved with the manufacturer's name: Utley's Patent #11.126–1908.
>
> On Monday, as the *Titanic*'s portholes slowly rose out of the water, we could gaze through them for a moment, as though we were looking through the liner's windows. We saw splashing waves where passenger rooms should be. That is what survivors, too, saw from their lifeboats on the morning of April 15, 1912. Those portholes looked out, in the early hours of that day, from five first-class staterooms on C-Deck–C-82, C-84, C-86, C-88, and part of C-90.
>
> Susan Wels

> Everyone wanted a souvenir from the *Titanic* expedition. The most unique were ones we created ourselves by sending down polystyrene cups with the submersible. Tied in an external bag, they were subjected to full pressure at two-and-a-half miles down. They came back as little miniatures with our names still legible. Cool! I thought my 9-year-old son would be pleased with this artifact and the idea of the crushing pressure that created it. But it was my second expedition to the site – I'd been gone two weeks – and the thought of me now disappearing into editing rooms and meetings for more months won out. "It sank! Get over it!" he sighed with exasperation. But some of us never do.
>
> Maureen Lemire

Museum people are often tasked with telling stories through third-hand information and piecing together what *really* happened from multiple sources. On the expedition I had a chance to meet everyone, really understand and document what they were seeking to learn, so we could produce the exhibition *Titanic Science*. The Discovery Channel work was great, but it was my task to take the science part of the stories and figure out how to create three-dimensional education experiences.

After returning to the Maryland Science Center, the exhibit team worked for over a year to bring all the expedition content to life. For an exhibit we named *Titanic Science*, we created sinking models, an interactive that attempted to avoid the iceberg, a close look at the rivets, an ROV model in a huge tank of water and a walk-through rusticle! The exhibit traveled throughout the US for about three or four years. Parts of it were then disassembled and used in other *Titanic* exhibits.

Stephanie Ratcliffe

I had the absolute honor and privilege to assist Dr. Roy Cullimore on the 1998 Titanic Expedition to perform some of the very first science experiments and examination of the biology and microbiology on the wreck. Twenty years ago gives some perspective to the expedition, of how quickly time passes, how young I was and the amazing experiences that I was able to have. I was able to spend some time on each of the four ships out there so each afforded me different perspectives of life at sea. The 1998 Titanic Expedition was my first but not last foray into the deep ocean. It has become a lifelong passion to explore the science of wrecks around the world, learning how the biology and microbiology is such an integral part of the ocean environment.

Lori Johnston

Being in the control rooms, with the images of *Titanic* from below on the monitors – fascinating. During my time out there, I met people with all kinds of expertise; rusticles, submersibles, passengers' biographies, naval architecture, etc. I really enjoyed those conversations.

Steven Biel

Many of the wrecks I work on are considered large by shipwreck standards when they are 50 feet off the ocean floor. *Titanic* is huge, and the altitude from sea-floor to superstructure was over 80 feet tall and just incredible. We are moving right along, and suddenly, this huge iron wall is in front of you. Straight up.

Troy Launay

I will always look back on *Titanic Live* in '98 as one of the greatest experiences and accomplishments of my life. I have since been back to *Titanic* twice – for another live show with James Cameron in 2005 and on the RMS Titanic Expedition in 2010. All were incredible experiences but '98 will always hold a special place in my heart.

Bob Sitrick

Dick told us and told everyone over and over how impressed he was with all the people he worked with on the expedition. He had great respect for all those involved, and was very happy to be a part of the project.

Beverly Silloway

Although I spent lots of time circling the tiny ship poking my nose in on all the scientists and historians at work, I also remember the Scrabble games and late-night talks in the galley. Many *Titanic Science* exhibit ideas were hatched during those times. Everyone was very generous with their time and tolerant of my endless questions. The exhibit could not have been created without everyone who helped develop the content. Today I am executive director of The Wild Center, a wonderful nature-based museum in the heart of Adirondack Park. It is a new museum, and I would love to be your host if you are ever in the mountains of upstate New York.

Stephanie Ratcliffe

The Scrabble game was brought on board *OV* at St. John's during the "intermission" between the expedition's Phase One and Phase Two, kindly supplied – in response to my query about where we might buy one while in the city – by the ship's agent who had "borrowed" it from his daughter. It provided many days and nights of enjoyment during Expedition '98. When the *OV* sank in July 2002, it was sad to think that this source of mental stimulation, diversion and occasional uproars about "legal" and "illegal" words now lay at the bottom of the Arabian Sea.

Charles A. Haas

When I was on this expedition, I was quite young. Having done many shipwreck expeditions, the most striking thing about *Titanic* is the undersea enormity of the wreck itself. I have been to *Titanic* several times over the years and it always amazes me. I hope that someday what is left of *Titanic* can been seen by all the world before it dissolves into only a memory.

Troy Launay

I owe a big debt to Tim Foecke for recommending me for the spot. After the voyage, Tim and I co-advised a PhD student named Jennifer Hooper who went on to write several papers, a dissertation, and a book with Tim Foecke called *What Really Sank the Titanic: New Forensic Discoveries*.

Tim Weihs

Dick had a small souvenir he picked up in Canada just before boarding the ship. It was a "Titanic Beer" with an image of the ship on the bottle. He had it above his desk. Though he passed away in 2010, the bottle still remained above his desk. Last year [2017], Hurricane Harvey hit us very hard, and we had tremendous damage at our home. The house was totaled. Of all the items we have not been able to find, that "Titanic beer" bottle is one that sticks out. I imagine it as being washed back out to sea – almost appropriate.

Beverly Silloway

In December 1998, model tests were performed at the Naval Warfare Center, Carderock, with a model of the *Titanic* bow to determine how it made its way to the seabed. Later calculations showed the rate of descent was between 22 and 28 knots. These tests were sponsored by the Discovery Channel and would be included in the Discovery Channel's documentary, *Titanic: Answers from the Abyss* with the assistance of Stardust Visual. In fact, David Elisco, the director of the documentary came to a meeting of the Marine Forensics Panel in January where the final script was examined for technical content.

Bill Garzke

Participating in the *Titanic* expedition is a career highlight. Being with everyone for those days at sea was invaluable.

Stephanie Ratcliffe

When you share an experience such as this, the people become friends for life and the memories are cast within you forever.

Troy Launay

Reflections

One of the questions Dick [Silloway] examined was, when looking at the bow, what caused the vast separation between it and the stern. He developed a theory that seemed very reasonable and plausible. It was decided to build a scale model to test the theory at the David Taylor Research Lab in D.C., and the test proved the theory was 100% wrong. "That's science!" Dick said.

Beverly Silloway

[Parapharased] I am exhausted and frustrated by the rumors over the years. People who have never been on expeditions create or spread rumors about the D-Deck door, the crow's nest, the mast light, holes in A Deck, the wall of Captain Smith's suite . . . . Can anyone give evidence supporting their rumors? Of course not. The *Titanic* is a wreck – a beautiful wreck – but a wreck. Step by step, the deterioration of the *Titanic* is more and more visible. In 2010, I saw that the height between the decks at the level of the grand staircase are now only six feet. In 1987 the height was between eight and nine feet. All the decks are collapsing on each other.

PH Nargeolet

The light towers' maximum autonomy was an hour and fifteen minutes. In order to be able to control the lighting timing, George Tulloch gave me a small barrette with four chronometers, one for each tower. I treasure it as a precious gift; she is always hung on the wall of my office, and when I look at it I think of this crazy expedition and this incredible lighting work of the *Titanic*. [These expeditions] were some of the most important adventures of my life. The thought comes to me of George, who left us. Without him and PH I would never have done this incredible work. Thank you, George. I think of you often.

Christian Petron

> Thank you, PH and George, for three experiences of a lifetime, for caring 24 hours a day; for constantly inspiring the very best from each of us, and for the knowledge you and our team uncovered and shared with all of us – and the world. May these golden memories – and so many others – last each of us a lifetime.
>
> Charles A. Haas

In 1999, George Tulloch was removed as President and CEO of RMS Titanic, Inc., in a hostile takeover. The new leaders of the company changed directions from the plans George had made and worked hard to accomplish. Those leaders are now gone. The Titanic Artifact Exhibitions still promote the artifacts, and have tremendous untapped potential. George Tulloch envisioned "standing in the corner at an exhibit, and watching young people walk up, looking at The Big Piece and learning about the *Titanic*'s story." He envisioned "creative new exhibitions as we learn more, and as technology takes us new places."

George talked about the "reflection room" of the exhibit. He wanted benches placed in the room with the list of victims written on backlit, acrylic wall displays. "I want people to be able to sit and reflect on all they've seen at the exhibit. I want them to see the names, and know that the *Titanic* story is about the people, not just the metal.

"You know something that really moves me, Billy? To see the emotion on people's faces [as they experience the exhibits]. It makes it all worth it."

George's dreams and hopes continue in many of us.

Stories from Expedition Members

Thoughts on the 1998 Expedition

George H. Tulloch, 1998 Expedition Co-Leader

We knew we had a good representation of artifacts from the wreck of *Titanic*. Dr. Eric Kentley of the National Maritime Museum of Great Britain had confirmed to us that the 5,000 objects recovered from 1987 through 1996 constituted a very powerful collection. So, as the summer of 1998 approached, our primary objectives were a follow-up on our ongoing science work and highly selective object recovery that would fulfill our responsibilities to the Norfolk Court as salvor-in-possession and legal guardian.

Our scientific work out at the wreck site would be our best effort to date. We again chose IFREMER, France's oceanographic institution, which had discovered the *Titanic*. Commander Paul-Henri Nargeolet was again in charge of all recovery work, as he has done throughout our research and recovery work. We again secured our relationship with the Discovery Channel, which would cover the expedition as it had in 1996. The Discovery Channel's interest in the science of *Titanic* was equal to ours, making a good foundation as we planned and prepared for the 1998 expedition. We recruited many of our past expeditions' science partners: Dr. Roy Cullimore, Canadian microbiologist from the University of Regina; David Livingstone, Harland & Wolff senior naval architect;

Thoughts on the Expedition

William Garzke, chairman of the Society of the Naval Architects and Marine Engineers' Forensic Panel; and Paul Matthias, chief scientist for Polaris Magnetic and Sonar Imaging and his staff. Our historical adviser team of Eaton, Haas and Wetterholm remained the same for the 1998 expedition, this team having covered all the recovery over the decade. Conservation was handled again by the on-board LP3 Conservation staff in their laboratory on *Nadir*.

To this research team we added Angus Best, oceanographic geologist for the University of Southampton, England; Pierre Valdy, chief technical engineer at IFREMER; Richard Silloway, marine forensic engineer; David Wood, naval engineer with Gibbs & Cox; Tim Weihs, material engineer for Johns Hopkins University; married physicists Susan and Bill Willard; Stephanie Ratcliffe of the Maryland Science Center and Greg Andorfer, its director.

We chose to use the same surface ships, the French Navy's *Nadir* and *Abeille Supporter*, and the *Ocean Voyager*. *Nautile*, the submersible mothered on *Nadir*, would be our workhorse, as always. We had brought two other unique submersibles in 1996, *Jules* and *Jim*, each a two-man acrylic sub, for shallow filming. They were mothered on *Ocean Voyager* that year. In 1998, we chose to leave them on land and use their space to develop a large remote-operated vehicle effort which would assist the crew in *Nautile* with a total of three robotic ROVs tethered from different ships on surface.

The expedition's goals would be fourfold:

- One: We would attempt the recovery of the large section of hull located in the debris field in 1994 by Commander Nargeolet and Dr. Kentley. We had studied the hull plating in 1996 and failed to recover it that year.

- Two: We would continue the ongoing biological, metallurgical and geological tests, and compile all of the data on site, under the eyes of the marine forensic team. It would be their responsibility to answer the tough questions of what happened and why, what is happening and why, and what will happen and why.

- Three: We would attempt to share this effort at sea with the world, in a live television transmission from the ocean surface, from the *Titanic* on the bottom, and from land-based studios, all at once.

- Four: We would then attempt to compile all our data, run it through the US Bureau of Standards and the National Transportation Safety Board and produce subsequent Discovery Channel programming that would serve the public's interest in *Titanic* knowledge.

We chose August 4 as our first workday on the site. We would keep this schedule thanks to work from hundreds, if not thousands, of caring people around the world. Discovery Channel, NBC News, IFREMER, Canal Plus, Oceaneering International, Travocean Submersibles, Classic Worldwide Productions, the Titanic International Society, Aqua +, LP3 Conservation, and TV Asahi are just a few of the institutions that were critical to this effort, and to the successes we achieved . . . .

I shared my hopes and dreams . . . with some of the finest people on earth, and I know what we accomplished for the *Titanic* story and for you . . . . I can . . . take comfort in my family and friends that made what we did a reality. I hope the objects from *Titanic* stay as a collection. If they do not, it will be a sad reflection on our culture, and another tragic moment in *Titanic*'s story.

Reprinted from *Voyage*, the official journal of the Titanic International Society, issue 33, Summer 2000, by kind permission of Cindy Tulloch and Titanic International Society.

Memories of *Titanic*

Sara James

I have been a journalist for many years and my job has taken me around the world. Sometimes the journey itself is unforgettable. Rocketing skyward in an F-15 E Strike Eagle in the Carolinas. Trekking in a four-wheel-drive from Peshawar to the Khyber Pass. Peering down at the Somali countryside from the door of a military helicopter flying from Baidoa to Mogadishu, less than a year before the terrible events relayed in the movie *Black Hawk Down*.

But the most extraordinary trip of my life took place in near total darkness. I will always remember the day I hitched a ride on a yellow submarine to the bottom of the ocean, to the wreck of the *Titanic*.

In our house in the hills next to the Wombat State Forest near Melbourne, Australia, I have the gift the crew gave me — a souvenir of that remarkable adventure.

Here's how it all began.

Back in 1998, NBC Dateline Executive Producer Neal Shapiro called me into his office to give me a dream assignment. I would co-host the Dateline/Discovery Titanic expedition special called *Raising the Titanic*. The entire broadcast would take place on the high seas, in the middle of the North Atlantic — a dazzlingly difficult, never-before-attempted technological feat.

Was I excited? You bet I was. I grew even more excited when I met all of the other remarkable men and women who were part of that expedition.

As the title of the broadcast suggested, the expedition intended to raise a large piece of the doomed ship. This section of the *Titanic* sat on the ocean floor, angled on a knife's edge, a short distance from the rest of the wreck.

Expedition organiser George Tulloch asked me to travel with the crew of the French sub *Nautile* as they tried to bring "The Big Piece" to the surface for the first time since that frosty April night in 1912. I said "yes" in a New York second.

The *Nautile* is many things. It's capable of plunging 20,000 feet beneath the waves. It's strong, with a spherical titanium hull. But it's not exactly roomy. Luckily, I'm not claustrophobic. And creature comforts are in short supply. There's no bathroom or heat. But who cares about luxury when the trip is to the bottom of the sea?

I can still picture the divers bobbing outside that sub, giving us a "thumbs up" that the hatches were secure. Suddenly the *Nautile* began to drop, and the ocean turned from blue to black as the light disappeared.

I talk pretty quickly at the best of times. As my nerves kicked in, I spoke at warp speed. I pulled out my camera and began to interview the crew, which helped me relax. They'd done this hundreds of times before, I figured.

When we got to the ocean floor, the two-man crew turned on the sub's floodlights, illuminating the jagged piece of the hull. I gasped in surprise as I saw a few dazed, strange looking sea creatures swim by. Then I heard the crew switch from speaking English to French, and I knew they'd reached the tricky part of their mission.

As the sub sidled over to The Big Piece, I peered out for a closer look. I realised I was looking through the porthole of the *Nautile* straight through a porthole of *Titanic*. I had a window to the past.

I felt a shiver pass through me as I thought about those who had travelled in that cabin, stared through that porthole. Suddenly, it all seemed too real. Who were they? What had happened to them? Had they survived, or perished? There were so many stories on that doomed ship.

For a moment, I confess that my casual confidence in the little sub wavered. Hadn't those on the *Titanic* also assumed "nothing could happen" to their state-of-the-art ship?

I took my mind off these thoughts by watching the crew intently. Using the robotic arms, they attached diesel-filled lift bags to The Big Piece. They moved carefully, cautiously. If they knocked the piece and it fell on the sub, we would have been trapped.

Once the lift bags were attached, we returned to the surface. At once point, the crew turned off the lights and I was treated to a dazzling shower of phosphorous as we ascended.

Back on deck, I was elated — and thrilled when they gave me a souvenir. That souvenir is a rock from the ocean bed near the wreck. A rock, from near the *Titanic*.

That rock travelled home with me on the *Petrel V*. On the airplane, back to New York. And when my Aussie husband and I and our two daughters moved to Australia, we brought that rock with us.

It sits next to our fireplace, and many a guest has hoisted it for a closer look.

The *Titanic* expedition was unforgettable. In addition, I have a wonderful rock to remind me.

From the Old World to the New

John P. Eaton

Those familiar with the details of *Titanic*'s legend and lore will recognize the heading as the title to a story by W.T. Stead that appeared in the 1893 Christmas issue of the *Review of Reviews,* of which he was the well-known editor. In it, the protagonists travel from the customs and manners of the Old World to the aspirations and enterprise of the New.

Somewhat prophetically, they travel across the Atlantic aboard a White Star liner named *Majestic;* in mid-Atlantic they rescue the survivors of a vessel, the *Ann and Jane,* that had sunk after striking an iceberg.

Stead himself would die aboard *Titanic* 19 years after his story appeared.

We do not look now to the coincidences of the story. We look to our departure from the past's colorful and traditional ways toward the future's high-tech science. We look to our own voyage, to our own transition from the old to the new . . . .

Thursday, July 30, 1998. Bayonne, France

After an unexpected and totally delightful two-day sojourn at the French seaside resort of Biarritz, our party of inquisitive seekers after truth depart the Playground of the Rich, via a fleet of French taxis. Fifteen miles and 12 minutes later we arrive at a flat and featureless quay at Base Navale de l'Adour, in the village of Anglet near the port city of Bayonne. Our dock is not in the city's bustling center. Rather, it lies on the River l'Adour, along a sparsely developed byway lined with occasional wood or cinder-block structures — bars, of course; storage sheds; a few clusters of habitable dwellings; an area that, if fog shrouded (which I am certain it frequently is), likely resembles an out-take from a Jean Renoir *film noir*.

Arriving shortly after noon, down the dusty road we see her from afar. The *Abeille Supporter* towers above the barges and fishing craft that line adjacent moorings. Yet the ship is not large, as such; the word that comes to mind regarding her appearance is "competent." We are greeted by the officers and crew; our luggage is taken from us and stowed aboard; we are told the vessel will not depart until after dark. We stand at dockside, shuffling our feet and wondering what to do for the next five or six hours.

Charles and I repair to the far side of the river, across a convenient bridge. We take photographs of the *Abeille* a quarter of a mile away. We return across the bridge. We stand on the quayside, alongside the *Abeille*.

Twenty-seven minutes have passed. The sun is still high in the sky. A nap will ruin tonight's anticipated sleep. We shuffle our feet on the dock's worn wood decking. A hot breeze lazily fans up the dust and blows it without enthusiasm across the amorphous landscape. Charles disappears. Six, perhaps seven hours to go before departure.

Ennui.

From afar the city beckons. A large cathedral's twin spires dominate the low skyline. A chapter from the past is waiting. Can we get to it, to examine it, to read it?

Another cloud of dust. This time it enfolds a car — a sparkling dark blue Citroën. The car emerges from the settling dust. A well-dressed lady and gentleman emerge. They introduce themselves:

Our Story

Monsieur *et* Madame Guilhem d'Elissagaray de Jaurgan. They have come in search of the lady's son, who is a member of the expedition. The son, however, is aboard another vessel of our *Titanic* search fleet that departed from another port and is already on its way to the wreck site. Disappointed, the mother sends her love. They prepare to go. M. de Jaurgan asks me if I would like to see the city and tells me they would be honored to show me around. Profuse with my thanks, I do not need a second invitation.

Bayonne is an ancient city in southwest France's Pyrenees-Atlantiques department. It lies on the River Adour near the Bay of Biscay and has a population of about 43,000.

In the Middle Ages it was held by the British, but has been occupied by France since 1451. It boasts a significant citadel, and its historic arsenal was famous during the 16th and 17th Centuries for the manufacture of cutlery and armaments; the bayonet was invented here (whence its name).

"You have been to Paris. You have just come from Biarritz," said M. de Jaurgan as we entered the city and passed along wide, flower- and tree-lined boulevards. "You have already seen what is Bayonne's New City. We will show you the Old City, and we think you will enjoy it."

We sweep up a curving road aside the stone parapet of the citadel. The road downward, toward the river, is narrower, more precipitous. We drive slowly. The buildings are stone and wood, smaller, older looking. The cobbled streets cause our heavy car to jounce. We turn onto a smoother street leading to a parking lot at river's edge. We park and prepare to climb the gentle hill toward the cathedral.

Across the sluggishly flowing Adour River lies Bayonne's "Old Town." Ancient wood and stone buildings in a sort of Mediterranean style line the river's bank. The buildings vary in width, the peeling paint on each either a scabrous white or a weathered pastel color — red, carmine, blue green, yellow — colors that perhaps 50 or 100 years ago might have been deep and brilliant. Very few of these houses are the same width; adjacent houses may be one to three feet wider or narrower than their neighbors. The ground floor of almost every building seems to hold a shop or a café, a bar or a restaurant. There appear to be many people, lots of activity along the waterside — quite cinematic. The houses' upper floors are shuttered; in the late afternoon heat, some shutters are open, many are closed.

We climb through a narrow shop-lined street that is scarcely 4 meters wide. The large cobblestones impede our footing until we learn to balance ourselves before and between steps. High above, past the shutters and parapets of buildings that were probably old in the time of the first Napoleon, a sliver of blue, blue sky is barely visible; it grants neither warmth nor light, only its own far-reachable presence.

Lining the narrow street, shops press closely against one another: butcher shops, pastry shops, food, candy, jewelry, wine, clothing shops; a fine, well-lit corner book shop. The shops offer no detraction or subtraction from the street's antiquated appearance. A step backward in Time has been taken.

The street opens onto a level place, a plaza atop the hill. Blue sky, trees, flower boxes, a gentle breeze; shops and outdoor cafés are active in the foreground. There are no vehicles; pedestrian traffic only. It appears to be the day's social hour; plaza, side streets, shops and cafés bustle with activity and points of color. Looming over all is the cathedral's tan and dark brown eastern façade . . . *La Cathedrale Sainte-Marie.*

Begun in the 13th Century, completed during the 15th and 16th Centuries and fully restored in the finest Gothic traditions during the late 19th Century, the twin-spired cathedral stands atop the gentle rise in the city's center. It transcends and commands the narrow streets, the plaza, the *people* so far beneath. It is approached and entered only after complete surrender to its awesome domination.

The ancient bishops and their architects planned well. Their cathedral is the home of an all-powerful God.

To enter the cloister, begun in the 14th Century, is indeed to enter the past. A shimmering stillness prevails. Ancient stone markers indicate final resting places for bishops, monsignieurs and priests; of donors to construction and restoration; to influential laity, all of whose bones and names are long dissolved to dust. Narrow, vaulted Gothic roofs have sheltered more than 30 generations of the faithful; the silence of their presence is gentle, yet persistent. Visible through the cloister's arched stone traceries, across the wide, verdant lawn, are the bases of the cathedral's west-facing twin spires. Staunch and steadfast, they define the cathedral's west end. Lighter stones among the darker show centuries of maintenance and repair to the soaring façades. Care has been thorough, and it has been loving. Atop the towers, delicately executed arching supports send the steeples heavenward.

To leave the cloister and enter the cathedral is as breathtaking as a sudden burst of thunder perceived yet not heard.

From the warmth of a late July afternoon to the cool, muffled echoes of time ago is but a step. But the breath catches at the vaulted ceiling's 86-foot height, the nave's 260-foot-long length and its 107-foot width. The faint aroma of incense lingers from the morning's Mass. So faint it can scarcely be heard is the present sound of a liturgical chant: vespers. My eyes scan around the cathedral as I search for the choir. It is a small group in white vestments, at the main altar more than 250 feet away. Their music is more than a millennium old; they sing in a church whose echoes are 600 years old.

I am humbled.

It is after 5 p.m., time to go. My host and hostess have a journey of more than 100 kilometers to travel home.

We turn right and traverse the cathedral's perimeter, past the large chapel devoted to St. Leon, the sacristy at the transcept's south extension, the high altar at our left, smaller chapels around the east end — the chapel to St. Anne, the Chapel of the Sacred Heart, St. Jacques, St. Joseph, St. Pierre, each with its own carved wood or marble altar, each with its garland or vase of fresh flowers. Saint Martin's Chapel adjoins the entrance/exit to the north porch. We turn left, pass through the massive wood doors and return to the terrace and steps that return us to the present.

Through the plaza with its cafés and shops, even more animated now, in the day's closing light. With electric lights, lanterns, candles, the scene is festively illuminated. I wish we could pause, sit at a table so that I could reflect on my emotions and what I have just experienced. But we must move along.

Turning left at the cathedral's southeast corner we walk the cobblestones back down the narrow street up which we had come so many centuries ago. We do not converse. The experience has moved all of us.

We reach the car and begin our short trip back to the dock at Anglet. As we reach the flats that lead to water's edge, I look back and over my right shoulder. The city is in silhouette against the scarlet, gold and purple sunset. Atop its low hill, the cathedral's massive structure stands out.

The twin spires are illuminated. A single star gleams in the blue, yet light sky.

My body will soon be on its way to the future.

I believe that part of my head shall always remain in the past, at *La Cathedrale Sainte-Marie* in Bayonne.

We arrive at dockside in the lowering twilight.

Our Story

Gratitudes are expressed, leaves are taken. My hosts depart, the car vanishes in the dusk. The air is still. A calm silence settles down, I take a last look behind me, toward the distant city, the illuminated cathedral. I board the boat and pass from the Old World to the New.

I step aboard a vessel whose metal decking and deck houses show signs of many years' use, but whose cranes, winches, generator housings, propelling and communications equipment are as clean and well-maintained as any aboard a Navy ship.

Abeille Supporter is described by its owner as: "[A] dynamic positioning, four-point mooring support vessel specially designed for diving maintenance, standby and submarine operations."

Built in 1975 by Aukra Bruk, Molde (Norway) Shipyard, it was converted to a diving support ship in 1976 at Ulstein Halto (Norway); it is owned by PROGEMAR and operated by Les Abeilles International.

Abeille Supporter would win no beauty contests. It is a functional vessel, built and operated to perform specific tasks. Almost one-third of its 209-ft length is the afterdeck, on which most of its work is performed; its overall breadth of 44 ft. gives it the stability needed to perform precision tasks involving great weights; it has a draft of 17 ft. The twin 12-cylinder, 4-stroke turbo-charged diesel engines, with a total output of 8,000 BHP, drive the vessel's nearly 1,038 gross tons at a speed of 14 knots.

Among its many hydraulic cranes and winches, the most prominent is the A-Frame, with its 30-ton lifting capacity, filling the entire width at the stern. The Brattwag hydraulic deck winch with a 30-ton lifting capacity complements the A-Frame's function.

Through an interfacing with *Abeille's* satellite navigating equipment, the dynamic positioning equipment, with the help of computers, enables the vessel to keep automatically in a set position with minimum modulation of propulsion equipment. This, in short, provides a stable "platform" — even in the middle of the ocean — from which accurate and exacting deep-water operations can be performed.

There is accommodation aboard *Abeille Supporter* for 42 officers, crew, diving personnel, technicians and — yes — passengers. The cabins, while far from spacious, are clean, well lighted, air conditioned and heated. The galley is fitted with splendid, modern equipment. The dining room is equipped with television, a VCR and a *most* interesting collection of tapes; many of these are in French, but for some reason it doesn't really seem to matter.

Abeille Supporter is not a holiday cruise ship. But for those engaged in pursuit of science and research, it provides comfort and security.

There is a hospital as well as two deck decompression chambers with six-man capacity.

The sun has set. It is quite dark, now, and hot. Along the quayside road, the village of Anglet is dimly illuminated. Music comes from the distant cafés. There is little motion, save the occasional passing truck or dusty car.

I stow my gear in the cabin I am sharing with Charlie and repair to the dining room for a cold supper. I am pleased to learn that *Abeille Supporter* is most assuredly *not* a "dry" ship — as are certain vessels in the flotilla that will assemble in mid-ocean a little more than a week from now. I am cheered by the thought.

The evening passes quickly. Our band of scientists, technicians, engineers, and TV production people, who have already shared three days together at Biarritz, come aboard one by one or in small groups. No one appears to have deserted.

Around 10 P.M. the gangway is pulled aboard. Mooring hawsers are loosened and hauled in. Without fanfare or sentiment we begin moving slowly down L'Adour toward the Bay of Biscay. Along our port beam the village lights flow and recede.

I am soon to be separated by time and distance from Bayonne and its cathedral. My thoughts are of the almost mystical communion with the Old World I experienced during my afternoon's visit. We are in the channel. The shore and its dim lights are far distant. Soon they will vanish. Soon there will be only the darkness of a French summer's night.

Downstream lies the Bay and the Atlantic Ocean. Ahead lies the dawn, the New World, the cutting edge of oceanographic science and technology.

Once again *Titanic* has worked its magic: Once again there is no Present — only Past and Future.

Dateline's Raising the Titanic and Dateline/Discovery Special Titanic Live

Bob McKeown with Meade Jorgensen

The Assignment

In the TV news business, interesting assignments fall into two main categories — the ones when you immediately call your wife and say, "Hey honey, guess where I'm going!" and the other ones, when you get a call from Human Resources asking for your next of kin because the network's high-risk life insurance policy is about to go into effect. When the assignment producer at *Dateline NBC* first told me about the Titanic Expedition, it seemed to be a bit of both.

Let me put it this way: I've covered about a dozen wars, hurricanes and tornadoes and I've come face to face with a polar bear in the Arctic and been bitten by a shark — on camera no less. But from the beginning it was obvious that this would be one of the most challenging television projects ever. First, there would be a highly-produced and tightly-edited hour about the Expedition for *Dateline NBC*. And almost immediately, that would be followed by a two-hour special for Discovery called *Titanic Live*, not only seen around the globe in over 100 countries but for the first time in television history, a live broadcast from the bottom of the North Atlantic at the final resting place of history's most legendary ship.

In addition to the 100 or so Expedition crew members, the combined *Dateline*/Discovery television production team numbered in the dozens. *Dateline* had two correspondents and three producers onsite, along with camera crews, video editors and transmission technicians out at sea. Editorially, we would plot the programs and gather the pictures and sound on location in the middle of the ocean. And back at NBC headquarters at Rockefeller Center in New York, there was a cast of literally hundreds more ready to catch our material, seamlessly put it together and get it on air, a failsafe to assure that whatever Mother Nature might have in store for us — bad weather, big waves, bigger icebergs — the *Dateline NBC* Special and *Titanic Live* would still be seen in North America and around the world.

That said, the potential problems of having the North Atlantic as our TV studio were too many to contemplate.

THE CHALLENGE

The Expedition leaders liked to say there's a reason humans got to the moon and to the top of Mount Everest before reaching the *Titanic*. So it was clear from the outset that technically, broadcasting the *Dateline NBC* Special *Raising the Titanic* and *Titanic Live* on Discovery Channel would be exceedingly difficult. Of course, we also knew the logistical issues of gathering material at the wreck site and getting it on the air — the unprecedented visuals of the world's most famous ship that our shows would depend upon — would be in the hands of the very best people in the field, indeed in the world.

Late in July, we visited RMS Titanic Expedition headquarters in Boston to meet those Expedition leaders, George Tulloch and PH

Nargeolet, for briefings on the technology we would rely on to do what had never been done before. First, the flotilla that would be the crucial technical platform 1,500 miles offshore — state-of-the-art ships *Ocean Voyager*, *Abeille Supporter* and *Nadir*. Next there were the manned and remotely-operated vehicles – *Nautile*, *Magellan* and *T-Rex* — to get cameras 2½ miles down to one of the most remote places on the ocean floor. And then the final piece, a revolutionary broadcast system sending those live images of *Titanic* back from the bottom of the Atlantic through cutting-edge fiber optic cable and on up to satellites that would — for the first time — transmit them to the NBC studios in New York, and ultimately around the world.

Out to Sea

As we were about to find out, not everything in the Expedition would be the state of the art. At the beginning of August, the flotilla departed from RMS Titanic headquarters in Boston bound for the North Atlantic wreck site. But there wasn't enough room on board for the entire *Dateline NBC* crew. So at about that same time we set sail from St. John's, Newfoundland, on a decidedly less high-tech vessel, a small Canadian cargo ship called the *Petrel 5*, which — politely put — was no *Titanic*.

As we wended our way eastward through the icebergs guarding St. John's harbour, it seemed we were heading back towards that infamous night in 1912 when *Titanic* was lost. Travel at sea, especially in a ship as venerable and barebones as the *Petrel 5*, is evocative of a different era. On the bounding main, time can pass slowly, now as then.

That is, unless you're linked to a satellite system that provides internet access and cellphone service which, as it happened, we were. For the 72 hours it would take to sail out to the *Titanic*, NBC producer Meade Jorgensen and I had work to do. As we reminded each other, a network television special and an historic two-hour live broadcast to millions of viewers around the world weren't going to produce themselves.

We began to assemble a to-do list for the *Dateline* and Discovery productions: which topics to focus on in each, what visuals and interviews would help us do it. In our tiny cabin, it was a claustrophobic exercise reminiscent of the subject at hand (think *Titanic* passengers in steerage). Fortunately, after a day at sea, the *Petrel 5* crossed into the warm waters of the Gulfstream and we could drag our research materials and briefing books up on deck, spreading out in the sun like college students cramming for final exams during spring break.

But the following day those Gulfstream breezes would fade and as we approached the *Titanic* wreck site, the winds out of the Arctic picked up and the waves began — exponentially — to grow, a weather pattern that would grip the North Atlantic for the next two weeks that we were there.

The four ships of the RMS Titanic Expedition — *Ocean Voyager*, *Abeille Supporter*, *Nadir* and our modest home away from home, *Petrel 5* — were anchored within a couple of square miles overhead the remains of the world's most famous ship, 12,000 feet below. It was from *Abeille* that the manned submersible *Nautile* was launched, the lynchpin for the missions down to *Titanic* and our live coverage of them. *Ocean Voyager* and *Nadir* were home base for the smaller remotely-operated vehicles — ROVs — called *Magellan* and *Robin*, about the size of a small car and a 27-inch TV set respectively, which could venture into confined spaces that *Nautile* couldn't — down the Grand Staircase where the cream of society came that night, through the wheelhouse where the *Titanic*'s master E. J. Smith so desperately tried to avoid the lethal iceberg and even hovering over the Captain's bathtub.

The Expedition staff of about 100 — scientists, historians, crew members — were based on those principal three vessels, while we

TV people were squeezed into tiny *Petrel 5*. Every day, the *Dateline NBC* crew would have breakfast in *Petrel's* cramped galley, then don foul weather gear and risk our lives to get from the wildly-rolling deck onto a 15-foot rubber runabout called a Zodiac, then shuttled to one or other of the larger Expedition centrepieces for shooting and interviews. And as the waves got ever bigger, our daily commute became more and more perilous.

A few years earlier, I'd had another "Honey, guess where I'm going?" assignment — to document a North Atlantic crossing aboard what was then the world's most luxurious ocean liner, *Queen Elizabeth II*. However, it just happened to be one of the roughest voyages in memory, with waves up to 35 feet high and a seasickness rate of about 100%. On the other hand, it was the perfect preparation for the Titanic Expedition because weathering 3-story waves on the little *Petrel 5* would be very different indeed from doing it on the 1,000-foot-long *QE II*. But after that experience, this time I knew enough to come fully equipped with pre-departure shots from the NBC medical department, time-release Dramamine patches and accu-pressure wrist bands to fight nausea.

All of which would be crucial because the weather kept getting worse, raising even more serious issues for us than dizziness and turbulent tummies.

If the seas didn't relent, there could be existential technical problems. For example, as the waves rose could the NBC satellite technicians still complete the link from the bottom of the ocean into space to facilitate the live broadcast around the world? My NBC colleague, producer Meade Jorgensen, who's seen it all in a decades-long TV career, says in his experience no technical feat has ever been better than the gyroscopic satellite dish that was lashed to the top of *Ocean Voyager* to keep the link to *Titanic* on an even keel, whatever the weather.

But as air time approached, the questions remained.

Could the crew still launch the ROVs 2½ miles down to *Titanic* to obtain those historic images? Especially under the crushing atmospheric pressure on the seabed, could the crew complete the link from the bottom of the ocean into space to facilitate the live broadcast around the world? And finally, could the *Nautile* be recovered in such dangerously high waves without risking the submersible and the lives of its crew?

We would soon find out.

AUGUST 12, 1998 – *DATELINE NBC* — *RAISING THE TITANIC*

Producing an hour of prime-time network television is demanding under any circumstances. Expectations are high, tempers can be short . . . and that's in the comfy confines of the *Dateline* studio and editing suites at Rockefeller Center in New York. Transpose all of that Type A adrenalin to the middle of the North Atlantic, getting tossed about by 35-foot waves and the degree of difficulty is multiplied exponentially.

First up, about a week after our arrival at the wreck site, was the *Dateline NBC* episode titled *Raising the Titanic*, focusing on the recovery of artifacts, in particular what is called the "Big Piece," a 17-ton section of the hull — almost 300 square feet — found near the *Titanic*'s resting place and the largest part of the ship ever retrieved.

A previous expedition had failed to raise The Big Piece when it was lost in stormy weather after being raised to within 200 feet of the surface. Now, two years later we were back to try again. Someone described the enormity of the task this way: "Raise the *Titanic*? Why not just lower the Atlantic?" And as the grappling technology on board *Abeille* struggled to haul that 17-ton section to the surface once again, the weather was getting even stormier. Given the history of The Big Piece, the moment of success — finally raised into the sunlight for the time in 86 years — was a very dramatic one.

That *Dateline* hour was recorded on video during our first week at sea, then packaged and aired from NBC News headquarters in New York on Wednesday, August 12. It featured those wondrous visuals of *Titanic* from the cameras aboard the submersibles, along with commentary from the people who knew more about the legendary ship than anyone else — naval architects . . . engineers . . . historians . . . technical experts . . . explorers. In a way, preparing that hour was like a *Titanic* boot camp, absorbing facts, figures and theories that would be crucial for the live Discovery special yet to come.

At a time when James Cameron's film *Titanic* was the hit movie of the year, *Raise the Titanic* was a smash on television too, giving *Dateline NBC* the best ratings in that timeslot not only that night but for the previous three months, no small feat in the ultra-competitive TV business, especially in the middle of the summer. But in hindsight, the real importance of that *Dateline* hour was as a dry run for the main event six days later on Sunday August 16, the historic two-hour live broadcast from the bottom of the ocean, produced by our *Dateline* team, hosted by Sara James and myself and shown on Discovery Channel around the world . . . *Titanic Live*.

AUGUST 16, 1998 . . . DISCOVERY CHANNEL . . . *TITANIC LIVE*

There is a unique feeling before a live television production . . . anticipation, nerves, sometimes fear . . . and especially before an unprecedented show like *Titanic Live*, to be broadcast from 8 to 10 P.M. in the Eastern Time Zone in North America and simultaneously, at all hours of the day and night, worldwide. And with recent conditions in the North Atlantic, there were substantial doubts whether we could pull this off.

The over-riding worry was that if the waves on August 16 were anywhere even close to the 30 or 35 feet we'd been seeing, they couldn't get the manned submersible *Nautile* — let alone the ROVs *Magellan*, *Robin* and *T-Rex* — safely into the water in the first place,

Our Story

in which case there would be no historic live visuals of the *Titanic* wreck and our unprecedented broadcast would be a failure. But the conditions could literally be a matter of life and death. Should the undersea fiber optic cables attaching *Nautile* to *Abeille* or *Magellan* to *Ocean Voyager* somehow become tangled with one another or with the massive *Titanic* wreck, it could literally be an existential threat. At 2½ miles down, there is no hope of rescue by anyone or anything. The tiniest crack in *Nautile's* hull and implosion would be instantaneous.

Just in TV terms, there was enough to worry about. Usually, live television — especially for a two-hour=long program — is done with an autocue or teleprompter system so the on-air personnel have the script close at hand. Follow the eyelines of the cast of *Saturday Night Live* and you'll soon see those humorous lines aren't just spontaneously rolling off their tongues; they're reading them. But because there were space and mobility issues in the cramped confines of the ships from which we would be broadcasting *Titanic Live*. it was simply impractical to install an autocue. So Sara James and I would essentially be working off the cuff, left to rely on a few scribbled notes, the prompting of our producers (thank you, Meade Jorgensen) and the facts and figures of the *Titanic* immersion course that we'd experienced since arriving at the wreck site.

And the scary possibility of two hours of international dead air was countered by the Expedition staff available to be interviewed by us, experts like naval architects Bill Garzke and David Livingstone; engineers Tom Dettweiler, Charles Haas and George Brotchi; *Titanic* historian Bill Willard; Expedition leaders George Tulloch and PH Nargeolet, and oceanographic imager Paul Matthias who was scheduled to be interpreting the *Titanic* wreck site from 2½ miles down inside *Nautile*. Whatever the weather, the only realistic danger of dead air from that all-star cast would be if they were . . . well, dead — which they most certainly were not.

But that Sunday August 16, as the clock ticked towards when we were scheduled to go live to the world, the waves outside our portholes were still roiling ferociously, no sign of relenting. If nothing changed — and for the past five days or so the conditions just kept getting worse — there would be no launch of the *Nautile* tonight and no historic live broadcast from the *Titanic* — obviously a problem for a project called *Titanic Live*.

But then, in late afternoon, almost incredibly, the seas began to change. *Dateline* producer Meade Jorgensen made sure to get a shot of the radar screen confirming the story we told — that there was a miraculous weather window calming those seas for precisely the period necessary to get our show on the air. And with just ten minutes to go, NBC's audio engineer Mike Noseworthy pressed the point by going up on deck and throwing in a fishing line, trolling the waters over the wreck site two and a half miles below, as if to say, "Weather? What weather?" Finally, we all could take a breath.

Indeed, by early evening the *Nautile* was in the water and the ROV *Magellan* followed. They were both 12,500 feet down adjacent to *Titanic* when we got the countdown from NBC headquarters in New York at 8 p.m. Eastern Time: "Three . . . two . . . one . . . ."

There's a picture of Meade Jorgensen which says it all. He's wearing seasickness bracelets on each wrist, along with two wristwatches, one for New York time, the other for local time in the North Atlantic. And he's holding a celebratory cigar — Cuban, from Newfoundland — to mark the successful completion of *Titanic Live*.

THE TRIP HOME

For the better part of three weeks, in the middle of the North Atlantic, angry seas looming around us for much of it, we could simply punch an NBC extension or a local New York number into our cellphones and speak to our colleagues or families as if we were at the office or at home. It created a false sense of security and the illusion that the *Titanic* wreck lies a lot closer to civilization than it actually does.

Then, a rude shock back to reality. As soon as *Titanic Live* was done, about midnight on August 16, our satellite link ended too. Internet access and cellphone service both came to a screeching halt. And the three-day return sail to St. John's, Newfoundland, became a journey back in time again. We'd experienced one of the most sophisticated peacetime operations ever at sea and now, much like when legendary writer Mark Twain travelled the world by steamship in the 19th Century, we had little else to do but read and think and reminisce about what we'd just been part of.

But as we left the North Atlantic, it was clear *Titanic* was already dramatically changing — even the wrecked bow section still so evocative of the magnificent ship that embarked on its maiden voyage, was poised to implode. Twenty years later, those haunting images we presented to the world are much of what remains to hold on to.

As our *Dateline NBC* colleague Sara James said at the end of *Titanic Live*, we'll remember because *Titanic*'s longitude and latitude are forever marked in our imaginations.

Memories

Bob Reid, Discovery Channel executive producer responsible for *Titanic Live*

I knew immediately a couple of things would be necessary to pull something like this off. One was, you have to be in a place and in a position so that the show could always go on regardless of circumstances. And the big circumstance was the unpredictability of the North Atlantic Ocean during hurricane season. What happens to the live coverage if a hurricane comes through? So right away, we knew we couldn't anchor the program from on location, on a ship at sea. If you're anchoring from a ship at sea, and you've got high winds, or severely adverse weather, and you can't broadcast, you can't go live; you're shut down. We had to lock down a place on shore to anchor the coverage. The idea was to have a foolproof anchor position, and a presenter who could take over and fill in if there were problems at sea; who could have backup content, so we could continue the live broadcast if there was a delay in the timing of expedition events, or the signal from the location of the *Titanic* was lost. That's why we worked with NBC News and *Dateline* and set up the on-shore studio at MSNBC (which was located in New Jersey at the time); it was the insurance, if you will, the guarantee that we could go forward with the live broadcast if the weather in the North Atlantic was a problem.

The guys who were at sea — they had the heavy lifting — literally and figuratively in terms of delivering the magical, once-in-a-lifetime, first-time-ever content. But the concept had to do with "live" and if it doesn't get on "live," it doesn't happen. So my role in all of this, along with the folks at MSNBC and all the other great producers and technical crew at the studio there, was to do anything and everything possible to make sure that the coverage of the expedition not only got on the broadcast and went live, but that it also went live as trouble-free as possible, and that the vision and the dream of the mission and the exploration was fully realized by being able to present the content in a seamless way so the worldwide audience could experience something it'd never seen before. Not only did you have this great fleet of ships out there, doing all the things the scientists and explorers had been doing over the weeks leading up to that day, but bringing a live picture of the *Titanic* from two-and-a-half miles down from a submarine — that was groundbreaking. I'm pretty certain that at the time, nothing like that had ever been done before.

Tethering that submarine to a two-and-a-half-mile-long fiber optic cable — and who knows what kind of danger that could have put the sub in — the point is, that was groundbreaking. All of it was groundbreaking and important. Our role on shore was to get that to the audience in as seamless a manner as possible.

When you ask me what the biggest headache was, it wasn't so much a headache, it was to be sure we had thought of everything we could think of, that we'd done everything we could do to make sure that we covered any possible contingencies.

One of the things we decided to do as a secondary kind of contingency was to do a complete run-through, a rehearsal, the day before; and I believe we set out to do it as close to the exact same time as the broadcast was scheduled to start the next day. So, the day before, we ran a live dress rehearsal. A very short time into the rehearsal, the satellite connection to the fleet went down because of weather and high winds. So we're thinking *"Oh boy, we're in trouble now,"* because part of what we were intending to do during that rehearsal was take in all the pre-packaged segments, and record live reports from all of the camera positions so we'd have all of those recorded and ready to go; even if we lost the live signal from the fleet during the actual broadcast, we could still present these pre-packaged segments. But when the rehearsal went down within minutes after we started — certainly during the first half hour (I don't remember how long we were going before we lost the signal) — then it was like, "Oh my goodness, what are we going to do now?"

We went to sleep that night not knowing if we were going to have something good enough to do the kind of broadcast everybody had worked so hard to prepare for the next day. When that signal went down, I think all the hearts in the control room stopped. We thought, *"What if we were live now? What would we do?"* Okay, sure, we could have had our anchor fill the time, and we could play the pre-recorded packaged segments, but it would not have been the same, and the entire mystique, the groundbreaking nature of the entire project would have been lost. Fortunately, the next day, we didn't have bad weather, or, at least not so bad that it shut us down, and we were able to do the program live, to bring the world *Titanic Live.*

The best of those live productions are like a symphony — everybody has to play together in key or it doesn't work. That's what makes live television so complex and also so special. There's a certain type of person who can do that (direct a show live) well, and you wash out pretty quickly if you can't handle it. You have to keep your cool when all about you is going crazy. It's knowing how, and being able to improvise, keep your cool, and be able to orchestrate a production live. Sometimes it's like a master musician's improvisation.

I have two favorite moments from the experience of doing the live program. The first one was when we saw the first live images of the *Titanic* from the cameras on the submarine, and I remember thinking, "This is truly historic." First, I was thinking of the technological accomplishment — and the danger — of bringing a live signal from a manned submarine hovering near the floor of the Atlantic Ocean, and watching the wreckage of the world's most famous ship. Then there was the fact that we could share that moment simultaneously with millions of viewers all over the world — live, at that very moment!

My second favorite moment was when the live broadcast ended, and we had gotten through it without a hitch, and the credits were rolling on the screen. As I watched the names of all the great people who made the expedition and the broadcast possible scroll by, I thought of the ambitiousness of the project itself, the vision, hard work, long hours and dedication it had taken to pull off one of history's great television events, and I felt very proud to have been part of it. And, I was especially happy that the broadcast had done credit to the vision of presenting the *Titanic* as never before seen, and that all the sacrifices and the risks had paid off in a great, nearly flawless, and historic broadcast. I was happy that despite the danger, no one had been hurt. I remember thinking, *"We did it. We pulled it off. We did something never done before in television history."*

TITANIC MEMORIES

Angus Best

As a young marine geoscientist who had recently joined the newly established Southampton Oceanography Centre (now known as the National Oceanography Centre, United Kingdom), I had little experience of seagoing expeditions. The previous year, I had spent 30 days on a research ship in the warm waters of the Arabian Gulf running a geological sampling and physical properties programme, far removed from the rough waters of the North Atlantic. The first I knew of the Titanic Expedition was when my Research Group Head asked me whether I was free over the summer to take part, apparently to help make a TV documentary. This sounded highly exciting to me, to have an opportunity to visit the actual *Titanic* wreck site. Of course, I was familiar with the *Titanic* story and its historic links to the city of Southampton.

I joined the *Ocean Voyager* at St. John's in Newfoundland and met the team, an eclectic mix of nationalities and accents (U.S., France, Canada, Sweden, UK . . .), mariners, business entrepreneurs, spouses, TV crew, marine architects, academics, historians, engineers, and scientists. The ship's mission was to deploy the *Magellan* ROV to collect close-up video footage of the wreck and assist various science projects, all to be filmed by the TV crew for the making of a documentary. This included, among other activities, using the *Magellan* ROV's robotic arms to collect samples of steel rivets for onboard metallurgy analysis, "rusticles" for microbiological studies, and geological samples for my own study.

One of the requirements of my joining the expedition was to devise an appropriate geological research study. After consulting with geotechnical engineering experts at the University of Southampton, I had established that it was possible to calculate the impact speed of the *Titanic* bow section from lab geotechnical tests on geological samples, and knowing how deeply the bow had buried itself into the seabed. All I had to do was obtain a few representative samples from around the bow section to take back to the Southampton labs.

In the meantime, I spent many fascinating days observing the others at work, viewing the video footage from *Magellan* as it explored the wreck, with expert comments from a retired USN submariner [Dick Silloway] and marine architects from the Harland and Wolff shipyard in Northern Ireland UK [David Livingstone], where *Titanic* was built and launched in 1911. The metallurgists (from the US) [Tim Weihs and David Wood] had set up a metal sectioning machine for slicing the rivets which they viewed under a microscope for flaws (to test the theory that the rivets were sub-standard, a contributory factor to the breach in the hull after the *Titanic* hit an iceberg on 14 April 1912). The biologists (from Canada) [Dr. Roy Cullimore, Lori Johnston] similarly studied "rusticles," like icicles but made of rusted iron, under a microscope to predict how long it would take the *Titanic* wreck to disintegrate. I remember taking a few screen tests with a TV cameraman, a soundman, the director and assistant all in a crowded lab space, where I attempted to explain my own science project.

The weather was calm most days, the food was passable, and I did not suffer any seasickness, of which I am susceptible, during two weeks of activity. There was always something new to observe on the *Magellan* video footage, like old leather suitcases lying on the seabed, or the jagged edges of pieces of wreckage, some intact sections still towering ghostly above the seabed. I recall being wary of half expecting to see a body suddenly emerge into view, but, thankfully, that never happened. I gave a few British TV and radio interviews on my return, and Radio Stoke (the home of the ill-fated Captain Smith) was concerned about the potential desecration of maritime graves. Indeed, the whole expedition was somewhat controversial because of the many artifacts recovered for exhibition to be held around the globe. It was open to criticism for "pseudo-archaeology," with a primary commercial motive, but that did not concern me at the time as my mission was scientific (and perhaps an amount of personal historical interest!). Of course, the profile of the expedition was enhanced by the recent release that year of James Cameron's highly successful movie *Titanic*, adding a romantic storyline to the tragedy of the *Titanic* sinking.

Finally, the *Magellan* team scheduled an attempt to collect a seafloor geological sample near the bow section. This involved inserting a plastic "drainpipe" into the seafloor using a robotic arm. Curiously, the whole seabed seemed to distort elastically downwards like pushing on a drum skin, and the corer would not penetrate. Several attempts proved unsuccessful, but eventually a grab sample was obtained instead by *Nautile*, the French manned submersible that was operating alongside from IFREMER's research vessel *Nadir*. The sediment sample was unceremoniously dumped in a bucket.

Quite suddenly, weather reports came in that a hurricane was coming our way, being [late August-early] September in the North Atlantic, the peak of hurricane season. For some reason, the *Ocean Voyager* was not heading back to St John's, but the *Nadir* was, so I had to make a boat transfer to the *Nadir*. It was a little hair raising stepping from the RIB to the *Nadir* as the sea was starting to get up, but I managed this safely. Initially, the *Nadir* seemed very pleasant. In true style, the French dining arrangements were excellent and I had a very pleasant lunch. I was amused by the notion that the Maitre d'hotel ranked above the Captain of the ship!

Worst of all, I found myself on a smaller vessel with a much livelier motion to get used to, plus a rising sea state — I rapidly became seasick. However, I stuck it out on the afterdeck long enough to see huge waves breaking over the superstructure before retreating below decks to my bunk. I don't remember much after that, apart from violent sickness and the incessant rolling and pitching of the ship, until we arrived at St. John's. I was land sick for four days while I waited for the next available flight home.

The geological sample study showed that the bow section probably hit the seabed at speed, as predicted from hydrographic modelling studies, rather than floating down "like a leaf falling from a tree," an alternate theory. The work was published in the journal Marine Georesources and Geotechnology: Best, A. I., W. Powrie, T. Hayward, and M. Barton (2000), Geotechnical investigation of the *Titanic* wreck site, *Marine Georesources and Geotechnology*, *18(4)*, 315-331.

Overall, the 1998 Titanic Expedition sticks in my memory as a somewhat random interlude in my early research career, but one I am grateful for nevertheless. The people and the stories they told as we dined in the mess, or drank coffee to stay awake on watch around the *Magellan* video screen, were all fascinating. I enjoyed this aspect of the expedition as much as anything else. My own TV clips did not make the final documentary, but I did show a rough VHS tape version to my children's class at school. That is one of the values and legacy of the 1998 Titanic Expedition — helping educate and inspire the next generation. I extend my thanks to all those I met on the expedition, for their comradeship and enlightening conversation, and lasting happy memories.

I still have as mementos some decorated "shrunken polystyrene cups" that had descended on the ROV to 2 miles water depth, crushed by the immense pressure down there, and some *Titanic* coal George Tulloch gave me in person.

Illuminating History

Christian Petron

These stories were submitted by Christian Petron in French and were translated and paraphrased for the reader.

I had been close friends with PH Nargeolet since our French Navy Divers' group days. As the expedition leader for RMS Titanic, Inc. in 1996, he contacted me to bid on an important component of that expedition. They wanted an expert in underwater systems to design and build a lighting system to shine on specific parts of *Titanic* that had never been seen before, and then later to illuminate the entire side of the wrecked bow.

I knew this would be an exciting technical challenge. *Titanic* was comparable to a six-story building, over 100 meters long, and at a depth of 3,840 meters. It was a unique and colossal challenge. I accepted!

PH, also an expert in underwater technology, suggested that I develop a battery-powered, self-sustaining project without cable feed from the surface. Because of the depth and the various currents at different depths, using a 4-km long cable didn't seem to be a good solution. For lighting, the system would use four towers, equipped with 5 HMI lights at 1200 W for a total power of 24000 W HMI, (the equivalent of 100,000 W conventional lamps).

For the buoyancy of the towers, I would use a technique involving balloons filled with incompressible, lighter-than-water liquid attached to the towers. Diesel fuel was the liquid of choice. The power for the lamps would be sealed lead gel batteries, such as motorcycle batteries. They would be put in a tank with oil, and the system sealed with a lid. I was familiar with this technique used in deep-water submersibles; it was part of my training as a submarine engineer in the French Navy. The wiring would be integrated into flexible hoses filled with oil so they wouldn't be crushed by pressure. I included three acoustic controls for each tower: one for powering it on or off, a second for geographical positioning, and a third to release the ballast chains so the towers could rise back to the surface once the batteries had run down.

There were dangers associated with the design. If the tower casings were to structurally fail and implode at 4000 meters, it would create a shockwave powerful enough to destroy a submersible. Designs had to be checked and rechecked. I submitted my design to PH and George, who shared it with Discovery Channel. Meanwhile, American and Canadian companies also responded to the call for bids; the whole idea was prestigious and could be used again after the *Titanic* expedition. My project, I was told later, was the only one with fully autonomous acoustic control towers; Discovery and RMSTI approved and accepted my proposal.

I was very happy, but couldn't rest on my laurels. Everything had to be constructed, delivered and ready for operation in four short months. It was quite a challenge because no one had ever built such material, but I was confident, and I absolutely wanted to succeed with this project. The towers were completed on schedule and were loaded onto *Ocean Voyager* to be taken to the *Titanic* site.

Once the teams arrived at the wreck site, the first task was to place a field of acoustic beacons all around the wreckage field, which would enable a submersible to accurately navigate to all the objects in the field. The purpose for the first dive with the towers was to illuminate the back, rarely seen part of *Titanic*, and

257

particularly the break where two large alternative machines could be seen, as tall as cathedrals.

I climbed into *Nautile,* and we began to descend. It took an hour to reach our intended depth. The submersible could dive for eight to ten hours at a time, so any trip down to *Titanic* could last up to 12 hours, with room for only the pilot, the navigator and one passenger/observer. I remember being very impressed the first time we dove in *Nautile.* As we peered out the portholes in more shallow water, everything was blue. As we descended further, we saw all gradient shades of blue, and by 450 meters, we were in absolute darkness. In order to relax everyone during the descent, the pilot played disco music, but he told us we couldn't dance in *Nautile.*

Our main objective was to locate and position each of the four towers. Each had dropped successfully; my calculations were correct in the weight of the towers, the volume of diesel needed in the balloon and the necessary weight of the chain for guiding the drop. Using *Nautile's* manipulator arms, we were able to position the towers and orient them for the first filming. This process took eight hours. The actual filming would be done another day.

Diving in a submersible produces a fatigue such that it's not safe to plunge on successive days. PH used *Nautile* the next day to install lift bags to The Big Piece for it to be raised to the surface. Once it was my turn to dive again, everything was ready and the order was given to light the towers. I remember how all of a sudden, there was a huge clarity as the lights illuminated the whole scene. It was so bright that the light reflected back through the portholes of *Nautile* and lit up her interior. Everyone in central operations was watching closely on their high-tech monitors, so they could see clearly any objects or movements on the sea floor.

We immediately started filming all around, and we could see an absolutely beautiful spectacle before us. The fracture with the two huge cylinders of the alternative machines were 20 meters high; everything was illuminated beautifully, bathed in a kind of surreal blue light on a black background. The towers had to be at least 40 meters away from the wreck in order to see the whole scene really well. We guided *Nautile* around this area and filmed it from all angles. The rear of the ship was totally unrecognizable; it was completely exploded. In fact, it didn't even appear to be in the shape of a ship, but more like a heap of scrap, completely nested in the bottom of the ocean.

Discussing it later with the naval architects who analyzed all the images, they explained to me that the stern of *Titanic* was extremely heavy and when it sank with the gigantic weight of the machines, it sank at an extreme speed and had to hit the bottom between 60 and 100 km per hour, so the hull was completely disintegrated. In fact, it sank so fast that, with its speed and the pressure of the water, the main deck turned exactly like a sardine box lid when it's opened with a key. The bridge was wrapped on itself and you could see the whole interior of *Titanic*.

When we got back to *Nadir* and opened the hatch of *Nautile*, the expedition team was there to congratulate us. They had witnessed our success. We watched the films immediately and celebrated over glasses of champagne. I was overjoyed because I knew I had passed this crazy challenge.

The towers were recovered to charge the batteries, check them out carefully and make any needed replacements of parts. The next dive would be for the purpose of lighting and filming the front part of *Titanic*. Areas were illuminated so scientists could analyze ferro-oxidizing bacteria that "eats" the metal and secretes forms of casting that look like stalactites. It is believed that since 1912, the hull has lost more than 30% of its metal.

Another problem that threatens *Titanic* is what Americans call "ice bombing." During October and November, the currents reverse and the cold Labrador Current passes over the area of *Titanic*. These currents bring icebergs, and underneath, there are sometimes huge rocks that break off and fall like bombs to the ocean floor. Under sonar, it's possible to see huge rocks that have absolutely nothing to do with the sand or debris around *Titanic*. In fact, near the ship, there is one rock that must weigh at least 80 tons.

On our dives to film the front part of the ship, we took magnificent shots, including travelling along the main first class deck located at the top of the wreck, finishing with a view of the front of the bow. We filmed the extreme front and all the way down the passage from the railing at the bow. We had left a tower in counter-light so we could see the whole stern. Discovery Channel loved these shots.

It was August 1996, and Cameron was still filming *Titanic*. I remember that when I saw the movie a year later, the scene with the characters Jack and Rose together on the bow brought back memories and the overwhelming emotion I felt when we executed the illuminated filming in '96.

The documentary films were broadcast during Christmas celebrations of 1996 in Washington and New York. They met with great success to the point that Discovery Channel immediately considered launching a second expedition in 1998. For this new excursion, my work was relatively simple: ensure the underwater filming only for the departure and return of *Nautile* for the *Titanic Live* production. The live show was a huge success, once again breaking records. I was glad to have participated.

Honored Friends

In the 20 years since the '98 expedition, the team members have taken many paths. For six weeks, they were a unique band of brothers and sisters, and afterwards, they went on to conquer new challenges and overcome new obstacles. Whatever their paths, a few of these individuals are remembered for their significant leadership qualities and contributions to the expedition.

PH Nargeolet, Expedition Leader

> ...a giant among men yet one of the most humble and decent people on Earth. Spending time with PH, you would never guess his incredible resume or experience.
>
> Bob Sitrick

One cannot talk about the successes of RMSTI from its beginning until 1999 without including PH Nargeolet in the conversation. PH planned and coordinated all of the expeditions and was instrumental as a liaison between the company and IFREMER. PH's integrity and professionalism made him the perfect person for the company George Tulloch wanted to build. In *Nautile*, PH has dived to the wreck numerous times. He has spent more hours on *Titanic* than the actual captain, E. J. Smith was on the ship. PH doesn't talk much about his dives unless you sit and talk to him. He remembers every corner, every passage, and every room. He is soft spoken when he tells how much the wreck has changed in the years he has been observing it. PH's experienced eyes, his technical and professional background, and his wisdom were just a few of the characteristics that created a strong bond of trust between George and PH. Whatever PH would recommend, George would do it. And PH was always correct.

Bill Willard

If George Tulloch was the engine of the three Research and Recovery Expeditions, his dear friend Paul-Henri Nargeolet was the rudder. To each expedition, PH brought his expertise in dealing with ships, the sea, submersibles and people. With a clear sense of purpose and calm deliberation, he saw "the big picture" and provided the organization and planning that ensured the expeditions' success. Despite that expertise, PH was – and is – a remarkably modest leader who, like George, gave all credit to "the team." His love for *Titanic* was evident in every decision he made. The endless hours he spent conferring with George and directing operations were exceeded only by his many dives to *Titanic*'s wreck. He was – and is – the world's true expert on that wreck, and he continues to offer invaluable guidance and remarkable insight into its condition, its future, and its exploration.

Charles A. Haas

TOM DETTWEILER, NAUTICOS

Overseeing all the underwater systems was a marine scientist and engineering expert named Tom Dettweiler. On this trip, Tom was making his own return to the *Titanic* after an absence of 12 years.

Tom was an underwater systems engineer on the 1985 French-American expedition that discovered the *Titanic*. The next year, he returned to the wreck with the second US expedition, this time shooting thousands of still photos of the *Titanic* with a camera-equipped underwater-towed device called *ANGUS*.

Tom knows the wreck so well, he admits it's as though he has a model of the *Titanic* in his head. "You can't ever get the *Titanic* out of your system completely," he concedes. So he's back in 1998, this time with far more sophisticated cameras, fiber-optic lines and the technology — for the first time — to broadcast live images from the manned submersible *Nautile*.

All the new technology, he predicts, will most likely lead to more mysteries about the *Titanic* and to more unanswered questions. "The *Titanic* is so big," he says, "that you could come out here 100 times and never learn everything about her."

Susan Wels

Tom Dettweiler told me about the *Jason Junior*. It had been lost off of the Ecuadorian coast and is resting at about 3,000 meters. It's regarded as lost although the US Navy had tried in vain to locate it. Nine vans of equipment, together with the barge, all was lost.

Claes-Gören Wetterholm

Pierre Valdy, IFREMER

If there is an engineering mastermind behind the '98 Titanic Expedition, it is Pierre Valdy. As the expedition's mission chief, he invented the novel strategies for recovering The Big Piece and achieving the live fiber-optic TV hookup from *Nautile*. "This year, there were two big breakthroughs," states Pierre, a project manager for IFREMER, France's oceanographic agency. "Technically, I think we've made a lot of progress."

A former motorcycle engineer in Paris, Pierre moved to Toulon on the southern coast of France and began designing submarines, *Robin*, *Nautile's* robotic eye, as well as special tools — such as a gentle suction pad

– to help *Nautile* recover a variety of objects from the *Titanic*'s debris field. Pierre also invented the technique of using lift bags filled with lighter-than-water diesel fuel to raise large objects, including The Big Piece, from the ocean. For much less delicate operations on far less famous and historic wrecks, Pierre invented another underwater tool, "The Grab" – a 50-ton remote-operated claw that can tear open a sunken ship like a tin can and bring up 200 tons of its contents to the surface.

"The Grab" has had some newsworthy success. The giant claw recovered 17 tons of silver coins from the *John Barry*, a US liberty ship that went down in the Sea of Oman, as well as 400 kilos of gold from the wreck of the *Douro*, off the coast of Lisbon. "On the *Titanic*, of course, you cannot use such a tool," Pierre declares. Instead, on this expedition, he has focused on more delicate challenges – improving the rigging of the lift bags in order to raise The Big Piece and reduce any possible danger to *Nautile*. And he came up with the long-shot plan for achieving the world's first live ocean-bottom TV link, 2.5 miles below the surface. "The success of both these projects," Pierre says, "was a bit of a surprise."

"I'm generally pessimistic about what I do," he explains, "because I always think about the problems. But maybe I should be more of an optimist – because till now, I have never made any big mistakes."

Susan Wels

Jack Bennett, NBC

. . . a true miracle worker – making the live broadcast a reality, running an incredibly complex operation, and remaining completely calm in the process.

Bob Sitrick

Troy Launay, Oceaneering

> ... a renaissance man if I have ever known one – fearless, incredibly talented, and a great kindness below a gruff exterior.
>
> Bob Sitrick

Mike Quattrone, Discovery Channel

> ... another incredibly inspiring yet decent man.
>
> Bob Sitrick

Greg Andorfer, Stardust Visuals

Greg Andorfer's role in the *Titanic* story is multi-faceted. In addition to the many documentaries for which he was the executive producer, he designed and created an exhibit — *Titanic Science* — that started at Maryland Science Center in Baltimore and toured the US. The *Titanic Science* exhibit was well received by attendees, and included hands-on displays, visuals, and even a mini-ROV driving exhibit where visitors could take the controls of a very small ROV and pilot it through a large water tank. In today's terms it was comparable to flying a small drone. Sadly, Greg has a medical situation that has taken much from him, including his memories. We salute Greg, and his contributions.

> Greg Andorfer ... ran the production side of things beautifully, captured the story impeccably, all while staying true to a sense of courtesy and decency.
>
> Bob Sitrick

Greg Andorfer is a person of many gifts and talents. People usually fall into either the left-brain (rational/scientist) or right-brain (intuitive/creative) category. But Greg is truly blessed with the exceptional capacity to fully utilize his entire brain. This is what made him the talented and award-winning producer of science documentaries, starting with Carl Sagan's *Cosmos* and ending with the *Titanic* series for the Discovery Channel.

Greg's love of life and love of unraveling the mysteries of life is only surpassed by his fierce and unconditional love for his children. Regarding his children, he often happily and proudly said, "My life is not my own."

Greg was diagnosed with early onset dementia over a decade ago. His precious God-given gifts of intellect, creativity, superior communication and expression have all steadily eroded, almost to the point of completely disappearing. Even so . . . the essence and "core" of Greg – his heart and his soul – still remains and will undoubtedly remain until his last day. Greg is a man who was always keenly aware that his life was not about him at all. Rather, he devoted his life to his children and the wonders of the endless universe around us.

Thank you, Greg Andorfer, for the life you have lived as an example for us all.

Charlene Haislip

Some members of the team are no longer living. The contributions they made were significant, not only for this project, but back in the "real world" after the conclusion of the expedition. To them, highest praise and commendation are offered for a job well done and for friendships over the years. It is regrettable they are unable to share in this 20th anniversary celebration. They are missed.

DAVID WARFORD, OCEANEERING

David was chosen for the Oceaneering team and was always known as the guy willing to get dirty to get the job done. Dave was always seen with a rag hanging out of his pocket, and he always looked like he had been working all day.

I only knew Dave Warford as we sailed together on a 2006 expedition. His nickname on board was "Goat."

Dave Jourdan

Dave was a key member on team *Magellan* and chosen for this expedition because he was the best sonar man in the business. He increased the overall team entertainment to new levels. At every safety drill it was Dave's signature move to arrive late with a new and unusual hat. He continued with Oceaneering for many years and went on to become a project leader there.

Troy Launay

Ron Schmidt, Oceaneering

Ron Schmidt could always be found somewhere near the *Magellan*. He would go from component to component, tightening bolts, adjusting connectors, cleaning lenses, calling for manipulator tests, and much more. He always had a small collection of tools with him, and cleaning cloths. He was a quiet man, but very knowledgeable.

I knew Ron Schmidt most of my life. We went to college together and became commercial divers at FIT. We were the best of friends. We worked together for many years and on many, many projects. Our families became very close over the years. All of my most valued offshore memories include Ron. It was very painful to me when he passed. He was laid to rest in the sea where he spent most of his life.

Troy Launay

George Brotchi, Oceaneering

George Brotchi was one the kindest, good-hearted men one could ever know. I worked with him for many years on numerous projects, and he would put up with a tremendous amount of foolhardiness from a very diverse cast of characters. That is one of the special things about him. Our code name for George was "Howdy Doody," and his smiling face would appear at some very unusual places when you least expected it. His guidance for Ron and me during the very intense photo mosaic portion of the project was critical.

Troy Launay

George often appeared to have a gruff exterior. Sitting with him for any length of time, one would come to realize that he was far from the initial appearance. His smile was contagious.

Bill Willard

DAVID LIVINGSTONE, HARLAND AND WOLFF

David was a member of the 1996 Titanic Expedition, and he's most interested in the technical aspects of the ship – how she was constructed and operated and the facts of how she sank. "From a professional naval architect's viewpoint," he says, "there's so much myth, legend and misrepresentation about the *Titanic*. I'm interested in determining the facts of what happened to the ship, because the facts are always much more interesting."

Susan Wels

DICK SILLOWAY, FORENSICS ANALYST

Dick Silloway was always one of the first to be at meetings, and at the viewings from the bottom at the site. He had a keen eye for noticing anything out of the ordinary as the cameras moved slowly across the wreck. Dick loved to ask questions in an attempt to learn the "*Titanic* history" and pair it with what he was observing. He was a part of many great discussions. Dick was retired Navy, and blended right in immediately with all of the rest of the team.

A few months after the expedition, Dick came to South Carolina for an event – "An Evening with *Titanic*." He, along with Bob Sitrick, Jack Eaton and Charlie Haas talked with more 125 guests in a wonderful event. Dick was awesome, bringing photos and slides to show and explain away. We had a wonderful weekend just enjoying conversations and telling stories.

Bill Willard

One aspect that really meant a lot to Dick was being involved in all the conversations with the others, where they discussed and shared ideas.

Beverly Silloway

George Tulloch, Expedition Leader, President RMS Titanic, Inc.

... one of the greatest men and friends I have known in my life. George was a true visionary and inspired those around him through word and deed. I miss him dearly.

I have many wonderful memories of George. One of my favorites is driving through France for meetings with IFREMER and the facility where they performed restorations on *Titanic* artifacts. George was driving in the fast lane but was not the fastest vehicle on the road – a French motorcycle cop pulled in front of us and at a very high rate of speed turned completely around backwards and gestured to us – I am assuming it meant move over or speed up.

Another [memory] that stands out was the night before the live show – we were expecting bad weather on live day, and there was a contingent who wanted to give up on the live portions and go with our rehearsal tapes. George was livid – banging an ashtray so hard on a tabletop that you could hear it below deck. He refused to give up on all the hard work of the past year and give in to fear. He won. The weather was too bad to launch the manned submersible right up until minutes before our deadline – then it miraculously cleared, the sub launched, and made it to the bottom with no time to spare.

My final memory was when George and PH traveled into the city to see my sister's art gallery debut – George was already pretty sick at that point but once more did not give in to fear or weakness – he spent a long time with us at the gallery and a restaurant afterwards. Shortly after that he passed.

Bob Sitrick

When George planned this thing, he could have left out the hurricane. But hey! Not everyone gets to have all the fun we had!

Anonymous

Friday, August 28. Midnight. After checking my email on the bridge, I went out onto the fantail. I saw George out there all by himself in the dark with a pack of Marlboros, a can of heavy duty Turtle Wax buffing compound, and an electric waxer, blasting an old Chuck Berry CD. He says that once every expedition he washes and waxes the submarine. "No one else," he says, "gives a shit about how the sub looks," but he does – it's a point of pride. So I helped him out for about half an hour. Truly one of the weirdest things I've ever done in my life. Midnight, in the middle of the freakin' North Atlantic, washing and waxing a yellow sub and rocking out to "Shake, Rattle and Roll."

Susan Wels

Honored Friends

Despite passage of two decades, my memories of this incredible expedition and the two expeditions that preceded it remain so very vivid. I cherish these memories, an enduring legacy of true highlights of my life. I often think about the amazing team we were and how well we worked together. I mourn the loss of dear expedition friends, and of course, George, over the past 20 years. And I remain deeply proud of all we accomplished in learning about history's most famous ship and literally rewriting its history.

Charles A. Haas

When I first met George Tulloch, I learned three things from him. First, the ship and her story would get the utmost respect at all times. This was primary. Second, the primary mission of the company RMSTI, was to preserve and exhibit the artifacts so that the story and true history of the ship would continue through generation after generation. Third, George trusted people to do what they said they would do. He treated everyone around him with respect. George took great abuse from some segments of the *Titanic* population that did not agree with salvaging the artifacts, yet George did not "counterattack." Instead he persevered and weathered the storms. He taught me many things during and after the 1998 expedition. I miss his counsel and his friendship deeply. Thank you George, for believing in us.

Bill Willard

Well done, George. Well done.

Anonymous

Author's Note

I owe a tremendous amount of gratitude to many people for making this project possible. Susan Wels immediately offered her daily posts from the site during the expedition. I used selected sections. Perhaps one day her entire work will be available online again so we can relive those times through her wonderful words.

I also owe a special thank you to my great friend Charles Haas. Charlie shared portions of his journal from the expedition, then synthesized the notes into descriptions that enable readers to visualize the event.

Many people sent their photos for us to consider. I wanted to include more than space would allow. To each of you, thank you.

I owe a special thank you to Daoping Bao, current president of Premier Exhibitions for not only supporting our project, but also allowing us to use the photos from the expedition.

It has been interesting to learn how the technology used on these expeditions has continued to make a significant contribution to international efforts. Meade Jorgensen shared a story that our satellite system uplink was used again a few years later during Operation Iraqi Freedom. If a satellite could be secured to a ship floating in the North Atlantic and send live reports from it, why not tow that satellite uplink system behind a military column during an armed conflict and give live reports there as well?

I attempted to contact every name on the expedition lists in my possession. There were numerous NBC specialists and Oceaneering crew members whom I was unable to locate. There were also members of the ship crews I sought, but was not successful. If you were one of our team members, send me some of your memories so we can include them in a possible second edition someday.

My friendship with George Tulloch continued after the expedition. We talked by phone nearly every day, sometimes more than once a day, beginning in late 1999. As I visit the artifact exhibitions, I always find a place from which to observe other patrons as they move through the displays. Strong emotions well up within me as I see their faces. In my heart, I know George would share the same feelings; in my mind, I feel he is right there beside me.

Ladies and gentlemen, it was an honor and a privilege working with you on the expedition. It is my hope you will cherish these collected stories and perspectives as I much as I have enjoyed reconnecting with you and compiling them into a complete story to share with others.

Bill Willard

Through more than 40 years of research, *Titanic* has played a major role in Bill Willard's life. Bill has conversed with some of *Titanic*'s survivors, studied and amassed a large personal library about *Titanic*, collected period memorabilia and commemorative items and more as he has further developed his knowledge about this world tragedy.

Bill is Instructor of Physics and Astronomy at UNC-Asheville and Asheville-Buncomb Technical College. For many years, he served public high schools in South Carolina, teaching physics, physical science, or astronomy. Hundreds of students over the years have benefited from his keen sense of understanding of how to communicate and teach.

Since the 1998 Expedition, Bill's research has focused on diverse topics such as the Rogue Expedition, the corporate operations of RMS Titanic, Inc., and multiple historical aspects of the families, individuals and crew on board the ship.

CPSIA information can be obtained
at www.ICGtesting.com
Printed in the USA
BVHW091001300722
643394BV00004B/9

9 781604 950410